The Moon: Humanity's Next Frontier

Daniel Fankhauser

ISBN: 9798346659662   Imprint: Independently published © scraibs.com. All rights reserved. This book, including its text, graphics, and overall design, is the original work of the author. Any content, structure, and creative expression not generated by AI is the exclusive intellectual property of the author and is protected under applicable copyright laws. No part of this book may be reproduced, distributed, or transmitted in any form or by any means, including photocopying, recording, or other electronic or mechanical methods, without the prior written permission of the author, except for brief quotations in reviews or articles. Disclaimer: Portions of this book have been generated using AI technology, specifically ChatGPT. While every effort has been made to ensure accuracy, some AI-generated material may not be entirely factual or reliable. Readers are encouraged to verify information from authoritative sources when necessary. The author assumes no responsibility for any inaccuracies or omissions arising from AI-generated content. All trademarks mentioned in this book are the property of their respective owners. The use of these trademarks is for reference purposes only and does not imply any affiliation or endorsement by the trademark owners. The author and publisher disclaim any liability for any loss or damage resulting from the use of information contained in this book. The content is for informational purposes only and does not constitute professional advice. The author asserts the moral right to be identified as the creator of this work and to protect the integrity of the work.

Contents

Prologue: Our Celestial Companion 4

Chapter 1: How the Moon Was Born 6

Chapter 2: The Science of the Moon 19

Chapter 3: The Moon Through History...34

Chapter 4: Moon's Role in the Solar System..................................42

Chapter 5: Exploring the Moon...53

Chapter 6: The Moon in Modern Science.......................................66

Chapter 7: Living and Working on the Moon 82

Chapter 8: Humanity's Next Great Economy 89

Chapter 9: The Future of Space Exploration 97

Chapter 10: Ethical Frontiers in Exploration 109

Epilogue: A Vision Beyond the Horizon....................................118

Prologue: Our Celestial Companion

For as long as humans have walked the Earth, the Moon has been a source of wonder and inspiration. It has lit the darkest nights, guided travelers across oceans, and served as a canvas for countless myths, dreams, and stories. Its presence in the sky, distant yet constant, has made it a companion to every generation—a silent witness to humanity's evolution.

From the ancient civilizations that charted its phases to the scientists who revealed its nature, the Moon has always been more than a celestial body. It has been a mirror for our aspirations, reflecting our endless curiosity and the audacity to reach for what lies beyond. Its surface, marked by billions of years of impacts, tells a story not just of the solar system but of Earth itself, offering clues about our shared origins and the forces that shaped our planet.

In the mid-20th century, the Moon became the stage for one of humanity's greatest achievements. When astronauts set foot on its surface, their steps symbolized the culmination of centuries of inquiry and imagination. Yet, the Apollo missions were not an end—they were a beginning, a reminder that the Moon is more than a trophy. It is a gateway to understanding the universe and ourselves.

Today, the Moon calls to us once more. Its icy craters and dusty plains hold the promise of discovery and the potential to support humanity's next great leap. Scientists and explorers see it not just as a destination but as a foundation for future exploration—a place where humanity can learn to live beyond Earth and prepare for journeys to Mars, asteroids, and beyond.

The Moon challenges us to think big and act boldly. It reminds us of the unity required to achieve the extraordinary, of the need to respect the environments we explore, and of the infinite possibilities that await when we venture into the unknown.

As humanity stands on the threshold of a new space age, the Moon remains our guide and partner. Its mysteries beckon, its resources tempt, and its presence inspires. This book is a celebration of the Moon's past, an exploration of its present, and a vision of its role in humanity's future. Through its pages, we will journey to the Moon and beyond, discovering not just what lies out there but what it means to be human.

The Moon is calling. Let us begin.

Chapter 1: How the Moon Was Born

The Moon, Earth's steadfast celestial companion, has captivated humanity for millennia. Its silvery glow punctuates the night sky, influencing the tides, inspiring mythologies, and serving as a cosmic beacon of mystery. However, the story of how this remarkable body came to exist is far older and far more complex than human history. The origins are intricately tied to the chaotic birth of our solar system and the dynamic interplay of forces that forged its enduring partnership with Earth.

To understand the formation, we must journey back 4.5 billion years to the dawn of the solar system. At that time, the Sun was a young star surrounded by a swirling disk of gas and dust—an environment scientists call the protoplanetary disk. This tumultuous region, filled with nascent planets colliding and coalescing, set the stage for the creation.

The Earth, too, was in its infancy. It formed from the accretion of smaller planetesimals—rocky fragments clumping together under the force of gravity. This process of planetary formation was far from gentle. The young Earth was a volatile, molten sphere, bombarded by meteors and heated by radioactive decay.

It was during this chaotic epoch that a cataclysmic event occurred, one that would forever alter Earth's destiny. This event, known as the giant impact hypothesis, is the most widely accepted explanation for the formation.

According to this theory, a Mars-sized body, often referred to as Theia, collided with the young Earth approximately 4.5 billion years ago. Theia's trajectory intersected with Earth's orbit, leading to an impact so immense that it liquefied both bodies' outer layers. This colossal collision ejected vast amounts of material into space, forming a dense ring of debris around Earth.

In the aftermath of the impact, gravity began its work. The debris—composed of material from both Earth and Theia—

coalesced over time, forming the Moon. This process likely took a few thousand years, a mere blink in geological terms. The newly formed Moon settled into orbit around Earth, its gravitational bond ensuring a partnership that would last billions of years.

The giant impact hypothesis is supported by compelling evidence gathered through lunar exploration and scientific analysis. One of the most striking pieces of evidence comes from the composition. Samples of lunar rocks, brought back by the Apollo missions, reveal a striking similarity to Earth's mantle material. This suggests that the Moon was formed from Earth's crust and upper mantle, rather than being a foreign object captured by Earth's gravity.

Additionally, the Moon lacks significant amounts of volatile elements—substances that vaporize at lower temperatures. This aligns with the idea that the formation involved extreme heat, as would be expected from a massive impact. Computer simulations of the giant impact scenario further support the theory, replicating size, orbit, and composition with remarkable accuracy.

While the giant impact hypothesis remains the most credible explanation, alternative theories have been proposed. One such theory suggests that the Moon formed alongside Earth from the same region of the protoplanetary disk. Known as the co-formation hypothesis, this idea posits that Earth and the Moon are "siblings" rather than "parent and child." However, this theory struggles to explain the unique composition and its lack of volatile elements.

Another hypothesis, the capture theory, proposes that the Moon was a wandering body captured by Earth's gravity. While this idea might explain certain characteristics, such as the irregular orbit, it fails to account for its chemical similarity to Earth.

Following its formation, the Moon underwent a dramatic evolution. The surface, initially molten due to the heat of formation, gradually cooled and solidified. This cooling process led to the differentiation of the interior, creating distinct layers: a crust, mantle, and core.

Evidence from lunar geology suggests that intense volcanic activity occurred during the early history. Lava flows created the dark basaltic plains we now call maria, visible to the naked eye as the "man in the moon" features. Over time, the geologic activity waned, leaving it a geologically "dead" world, its surface shaped primarily by impacts from asteroids and comets.

Long before science provided answers, humanity sought to understand the Moon through stories and myths. Ancient cultures across the globe wove the Moon into their creation myths, often associating it with powerful deities or cosmic events. In many traditions, the Moon was seen as a symbol of femininity, fertility, or the passage of time.

For example, in Greek mythology, the Moon was personified as Selene, a goddess who rode across the night sky in a silver chariot. In Chinese legend, the Moon is home to Chang'e, a goddess who fled there after drinking an elixir of immortality. Indigenous peoples of the Americas often linked the Moon to stories of creation and cosmic balance.

While these tales differ vastly in detail, they reflect a universal human desire to connect with the Moon and find meaning in its presence. Modern scientific understanding of the Moon began with the advent of telescopes. Galileo Galilei was among the first to observe the Moon in detail, sketching its craters and mountains in the early 17th century. His observations marked a turning point, revealing that the Moon was a rugged, imperfect world, not the smooth, divine orb many had imagined.

The 20th century ushered in a new era of lunar science with the Space Race. The Apollo missions of the 1960s and 1970s provided an unprecedented wealth of data, transforming our understanding of the origins and evolution. These missions confirmed the giant impact hypothesis and highlighted the potential as a key to understanding the history of the solar system. The formation of the Moon was not just a dramatic event in the history of the solar system; it profoundly shaped the evolution of Earth itself. The newly formed Moon began exerting its gravitational pull on Earth, stabilizing the planet's axial tilt. This stabilization played a critical role in creating a relatively consistent climate over geological timescales, allowing life to flourish on Earth.

Without the Moon, Earth's axial tilt would wobble chaotically, leading to extreme climatic shifts that might have hindered the development of complex life. For this reason, some scientists argue that the Moon is an essential component of Earth's habitability—a unique partnership that sets our planet apart in the cosmos.

In the immediate aftermath of the giant impact, the Earth-Moon system looked vastly different from what we know today. The Moon orbited much closer to Earth, appearing several times larger in the sky. Its proximity created immense tidal forces, generating enormous waves in Earth's molten surface and oceans. These tides, far more extreme than those we experience today, contributed to the mixing of Earth's early oceans and may have influenced the origin of life.

Over billions of years, the orbit gradually expanded due to a process known as tidal friction. As the Moon pulls on Earth's oceans, it creates a slight bulge that lags behind the orbit. This lag causes energy to transfer from Earth's rotation to the orbit, slowing Earth's rotation and pushing the Moon farther away. Even today, the Moon continues to drift away from Earth at a rate of about 3.8 centimeters per year—a reminder of the dynamic relationship between these two celestial bodies. The formation offers invaluable insights into the broader processes of planetary evolution. By studying the Moon, scientists can reconstruct the conditions of the early solar system and understand the mechanisms of planetary differentiation, impact cratering, and volcanic activity.

The giant impact hypothesis, for instance, underscores the chaotic nature of planet formation, where collisions were commonplace. Such events may have been critical in shaping not only the Earth-Moon system but also other planets and their satellites. For example, the retrograde orbit of Neptune's moon Triton suggests a similarly dramatic origin, potentially involving a collision or capture event.

Unlike Earth, the Moon has no atmosphere or significant geological activity to erase its ancient surface features. This preservation makes it a unique natural laboratory for studying the history of the solar system. The craters, formed by billions of years of impacts, provide a record of the intensity and frequency of meteorite bombardments. By analyzing these craters, scientists can infer the timing of major events in the solar system's evolution.

Moreover, the lack of atmosphere has allowed it to retain volatile compounds in its shadowed craters, such as water ice. These compounds, discovered in recent missions, are remnants of cometary impacts and solar wind interactions, offering clues to the distribution of water and other essential materials in the early solar system. The regular cycles—waxing and waning in the night sky—have made it a natural timekeeper for humans throughout history. Early civilizations relied on the Moon to track months and seasons, with many ancient calendars based on lunar phases. This practice underscores the profound influence on human culture and its role as a bridge between the natural world and human understanding.

The word "month" itself derives from the Moon, reflecting its significance in dividing time. For millennia, the lunar cycle was a cornerstone of agricultural, religious, and societal planning, demonstrating the enduring presence in the rhythms of human life.

While the giant impact hypothesis is widely accepted, certain aspects of the formation remain enigmatic. For example, the isotopic similarity between Earth and the Moon—particularly in oxygen isotopes—is strikingly close. This resemblance suggests that the Moon formed almost entirely from Earth's material, yet computer models indicate that a significant portion of the Moon should originate from Theia. Reconciling these observations is an active area of research.

Some scientists propose that Theia's material was thoroughly mixed with Earth's during the impact, creating the observed isotopic homogeneity. Others suggest alternative scenarios, such as multiple impacts or a larger, vaporized proto-Moon that eventually condensed. These ideas highlight the complexity of planetary formation and the ongoing quest to refine our understanding of the origins.

The formation has implications not only for Earth's history but also for the search for life elsewhere in the universe. By studying the Earth-Moon system, scientists can identify key factors that make a planet habitable, such as stable climates and the presence of water. These insights guide the search for exoplanets with similar conditions, expanding the scope of astrobiology.

Additionally, the Moon itself could harbor clues to the origin of life. While it lacks the conditions necessary for life as we know it, its ancient rocks may contain records of early solar system chemistry. These records could reveal the building blocks of life or the processes that shaped prebiotic environments on Earth and other planets.

Beyond its scientific importance, the Moon has been a source of inspiration and wonder throughout human history. Its ethereal presence has sparked countless works of art, literature, and music, from ancient hymns to modern poetry. The phases, representing cycles of growth, decline, and renewal, resonate deeply with human experiences, making it a powerful symbol in cultures worldwide.

In many ways, the Moon serves as a mirror, reflecting humanity's aspirations and curiosities. Its surface, marked by impacts and time, is a reminder of resilience and endurance—a celestial testament to the passage of billions of years.

The gravitational influence has been a critical driver of Earth's geological and environmental processes. Beyond its role in stabilizing Earth's axial tilt, the Moon has shaped the rhythm of our planet through tidal forces. These forces are not limited to the rise and fall of ocean waters; they also affect Earth's crust and mantle, subtly influencing tectonic activity.

In the early Earth-Moon system, when the Moon was closer to Earth, tidal forces were far stronger. These intense interactions may have played a role in the formation of Earth's early continental plates. The constant kneading of Earth's surface by lunar tides could have promoted the fracturing and recycling of crustal material, laying the groundwork for plate tectonics—a process crucial to Earth's ability to support life.

Moreover, the presence helped to mitigate extreme orbital variations that could have led to more erratic climates. By providing a stabilizing force, the Moon contributed to the long-term climatic stability that allowed life to emerge and diversify.

The rhythmic tides driven by the Moon may have been a key factor in the origin of life on Earth. Early life likely originated in shallow tidal pools or along coastal margins, environments that experienced regular cycles of wetting and drying due to tides. These cycles could have concentrated organic molecules, promoting the chemical reactions necessary for the emergence of life.

The influence extends beyond its physical effects. Its phases and cycles became embedded in the behaviors of early organisms. Many marine species still synchronize their reproductive cycles with lunar phases, a testament to the enduring role as a natural timekeeper.

The Moon serves as a preserved archive of the solar system's history, offering insights that Earth's dynamic surface has erased. Unlike Earth, where plate tectonics and erosion continually reshape the surface, the surface has remained relatively unchanged for billions of years. As a result, the Moon retains a pristine record of early solar system events, including the period of heavy bombardment—a time when the inner planets were pummeled by asteroids and comets.

This bombardment played a significant role in shaping the Earth-Moon system. On Earth, these impacts may have delivered water and organic materials, contributing to the conditions necessary for life. On the Moon, they created the craters and basins we see today, providing a timeline of solar system evolution.

Despite centuries of observation and decades of exploration, the Moon continues to harbor mysteries. One such enigma is the far side of the Moon, which remains perpetually hidden from Earth due to tidal locking. While robotic missions have provided detailed maps of this hemisphere, it remains a region of intrigue, with its thicker crust and unique composition raising questions about the early history.

Another puzzle is the presence of lunar water. Initially thought to be entirely dry, the Moon has been revealed to harbor water ice in its permanently shadowed craters. This discovery has profound implications for both science and future exploration. Scientists are still unraveling how this water arrived—whether through cometary impacts, solar wind interactions, or internal processes.

As the closest celestial body to Earth, the Moon has been humanity's first stepping stone into the cosmos. Its formation story is not just a tale of our natural satellite but a key to understanding planetary systems throughout the universe. The lessons learned from the origin and evolution have guided the search for exoplanets and their moons, expanding our knowledge of potentially habitable worlds.

For example, the concept of a stabilizing moon has informed criteria for assessing exoplanet habitability. By studying the Earth-Moon system, scientists have identified key factors—such as axial tilt stability and tidal influences—that could support life on distant planets.

The formation and its subsequent role in Earth's history have created a profound connection between humanity and its celestial neighbor. This relationship is reflected not only in science but also in art, mythology, and culture. From ancient rituals aligned with lunar cycles to modern-day lunar exploration, the Moon has been a source of inspiration and discovery.

The Apollo program, which marked humanity's first physical contact with another world, was a milestone that underscored the significance. The samples returned from the Moon provided direct evidence of its origin, confirming aspects of the giant impact hypothesis and revealing its shared history with Earth.

As technology advances, scientists continue to refine their understanding of the formation. Missions such as NASA's Artemis program and China's Chang'e series aim to return humans and robots to the Moon, where they will gather new data to address lingering questions. Advanced instruments and techniques will allow researchers to study the isotopic composition, internal structure, and surface features with unprecedented precision.

These efforts will not only deepen our understanding of the origin but also provide insights into broader cosmic questions. How common are Earth-Moon systems in the universe? What role do moons play in the development of life? The answers to these questions may redefine our understanding of planetary systems and the conditions necessary for life.

The story of how the Moon was born is a testament to the interconnectedness of cosmic events. From the chaos of the early solar system to its current role as Earth's steady companion, the Moon has been an integral part of our planet's history and future. Its formation highlights the dynamic processes that shape planetary systems and underscores the profound impact of celestial bodies on the evolution of life.

As humanity looks to the stars, the Moon remains a symbol of our cosmic origins and our boundless curiosity. It is both a witness to Earth's past and a guide to its future, a reminder that the answers to some of our greatest questions may lie just beyond our reach.

The gravitational relationship with Earth extends far beyond its immediate effects on tides and axial stability. This cosmic partnership has had profound implications for Earth's geological and biological history, shaping not only the planet's climate but also its capacity to host and sustain life. Its presence has acted as a stabilizing force, preventing wild oscillations in Earth's axial tilt, which would have drastically altered the climate over geological timescales.

Without the influence, Earth might have experienced extreme swings in axial tilt, leading to periods of severe glaciation or scorching heat. Such erratic climatic conditions would have made it difficult for complex life to evolve, much less thrive. Instead, the steadying effect has fostered a relatively temperate environment, allowing ecosystems to adapt and diversify over millions of years.

The influence extends into the rhythms of life itself. Tidal forces, driven by the gravitational pull, have played a role in shaping the biological clocks of countless species. These tidal cycles influenced the behavior and reproduction of marine organisms long before humans appeared on the scene. Even today, many coastal species synchronize their activities with the ebb and flow of the tides, a legacy of the enduring impact on life.

In the evolutionary narrative, the cycles may have provided early life forms with a predictable environmental pattern. Regular tidal movements concentrated nutrients in coastal areas, creating fertile zones where primitive organisms could thrive. These nutrient-rich environments may have served as cradles for the development of multicellular life, accelerating the complexity of Earth's biosphere.

As humanity's closest celestial neighbor, the Moon provides a unique opportunity to study the history of the solar system. Its surface, untouched by the weathering forces of wind and water, preserves a nearly pristine record of ancient events. The craters and basins scattered across the face tell a story of intense bombardment during the early solar system, a period when collisions between celestial bodies were a common occurrence.

One of the most significant periods recorded on the Moon is the Late Heavy Bombardment, which occurred approximately 4 billion years ago. During this time, the inner planets—including Earth—experienced a surge of asteroid and comet impacts. These collisions played a crucial role in delivering water and organic molecules to Earth, potentially setting the stage for the emergence of life. By studying the surface, scientists can piece together a timeline of these events, shedding light on the processes that shaped our planet's early environment.

Despite decades of exploration, many aspects of the history remain shrouded in mystery. For instance, the far side of the Moon—perpetually hidden from Earth—presents a vastly different landscape from the near side. The far side is dominated by rugged highlands and is sparsely populated with the dark basaltic plains, or maria, that characterize the near side. This asymmetry raises questions about the formation and the internal processes that shaped its evolution.

Recent discoveries have also highlighted the presence of water on the Moon. While initially believed to be a barren, arid world, the Moon is now known to contain water ice in its permanently shadowed craters. These regions, located near the poles, are some of the coldest places in the solar system. The discovery of water has profound implications for understanding the history and for future exploration, as it could provide a vital resource for sustaining human missions.

The formation and evolution offer insights not only into our own planet but also into the broader dynamics of planetary systems. By studying the Earth-Moon system, scientists have developed models to explain the behavior of moons and their host planets throughout the cosmos. For example, the processes that led to the formation via a giant impact may also explain the origins of other large moons, such as Jupiter's Ganymede and Saturn's Titan.

The lack of atmosphere and tectonic activity makes it an ideal reference point for studying impact cratering and surface evolution. These features, preserved for billions of years, provide a baseline for understanding similar processes on other celestial bodies, from Mercury's scarred surface to the icy moons of the outer planets. The journey from its chaotic birth to its current role as Earth's steadfast companion is a tale that mirrors humanity's own quest for understanding. Throughout history, the Moon has been a source of wonder and inspiration, a constant in the ever-changing night sky. Its cycles have guided ancient farmers, inspired poets, and spurred the imaginations of scientists and explorers.

The allure lies not only in its beauty but also in its mystery. It is a world both familiar and alien, a place that feels within reach yet remains largely unexplored. This duality has made the Moon a symbol of human curiosity and resilience—a reminder that even the most distant horizons can be conquered with ingenuity and determination.

As we continue to unravel the origins, we stand on the brink of a new era of exploration. Missions like NASA's Artemis program aim to return humans to the lunar surface, this time with the goal of establishing a sustainable presence. These endeavors are not just about revisiting past achievements but about forging a new relationship with the Moon—one that integrates science, technology, and humanity's enduring spirit of discovery.

The role in the future of space exploration is undeniable. As a stepping stone to Mars and beyond, the Moon will serve as a testing ground for the technologies and strategies needed for interplanetary travel. It will also provide invaluable resources, from water for life support to materials for building habitats and fuel depots.

The story of the formation is ultimately a story of connection. From the fiery collision that gave birth to the Earth-Moon system to the delicate gravitational dance that has shaped their shared history, the Moon and Earth are inextricably linked. This bond has defined the trajectory of life on our planet and will continue to influence humanity's future as we reach for the stars.

The Moon reminds us of our place in the universe—both our origins and our potential. It is a symbol of the profound interconnectedness of cosmic events and a testament to the enduring power of exploration and discovery.

Chapter 2: The Science of the Moon

The Moon, Earth's closest celestial neighbor, is a world of striking contrasts and enduring mysteries. From its stark landscapes to its intricate relationship with Earth, the Moon has captivated scientists for centuries. As the most studied celestial body outside of Earth, it serves as a cornerstone of planetary science, offering invaluable insights into the processes that shaped the solar system.

Layers and Composition

The structure is composed of distinct layers, much like Earth. Beneath its barren surface lies a story of ancient geological processes that have defined its appearance and properties.

1. **The Crust:**
   The outermost layer, the crust, is primarily composed of silicate rocks. This layer varies in thickness, averaging about 50 kilometers on the near side and up to 100 kilometers on the far side. The crust is covered by a fine, powdery layer of regolith—an accumulation of dust and rocky debris created by billions of years of impacts.

   The regolith holds a record of the history, preserving evidence of past meteorite impacts and solar wind interactions. It also contains valuable resources, such as oxygen bound in minerals and traces of water ice, which have sparked interest in its potential for supporting future lunar missions.

2. **The Mantle:**
   Beneath the crust lies the mantle, a layer of solid rock rich in minerals such as olivine and pyroxene. The mantle played a critical role in the early volcanic activity, which gave rise to the dark basaltic plains known as maria.

These vast plains, visible from Earth, are remnants of ancient lava flows that occurred billions of years ago.

3. **The Core:**

   At the center lies a small, partially molten core composed of iron and nickel. Unlike Earth, the Moon lacks a strong magnetic field, suggesting that its core is relatively inactive. However, traces of magnetism found in lunar rocks indicate that the Moon may have once had a more dynamic core, capable of generating a magnetic field.

The surface is a patchwork of craters, mountains, and plains, each feature bearing the scars of a tumultuous past. Impact cratering is the dominant geological process on the Moon, with countless craters ranging from tiny pits to massive basins. The surface can be divided into two main regions: the maria and the highlands. The maria, which make up about 16% of the surface, are relatively smooth plains formed by ancient volcanic activity. In contrast, the highlands are rugged, heavily cratered regions that date back to the earliest history.

The difference in appearance between the maria and the highlands is due to their composition and age. The maria are rich in basalt, a dark volcanic rock, while the highlands are composed of anorthosite, a lighter-colored rock that formed as the crust solidified. The craters provide a window into its history. Large craters, such as Tycho and Copernicus, are surrounded by bright rays of ejected material, indicating relatively recent impacts. Older craters have been eroded by subsequent impacts and are less distinct.

The largest impact features on the Moon are the basins, such as the South Pole–Aitken Basin, which spans over 2,500 kilometers in diameter. These basins reveal clues about the internal structure and the intensity of the bombardment that shaped the early solar system.

The gravitational pull is responsible for one of Earth's most familiar natural phenomena: tides. By pulling on Earth's oceans, the Moon creates bulges of water that result in high and low tides. This interaction is a reminder of the close relationship between Earth and its lunar companion. Beyond its effects on tides, the gravity has influenced Earth's rotation and orbital dynamics. Over time, the gravitational pull has slowed Earth's rotation, lengthening the day and stabilizing the planet's axial tilt. This stabilizing effect has been crucial for maintaining a relatively stable climate, supporting the development of life. The phases, from new to full and back again, are a result of its changing position relative to Earth and the Sun. As the Moon orbits Earth, sunlight illuminates different portions of its surface, creating the familiar cycle of phases that repeats every 29.5 days.

Eclipses occur when the Moon aligns with Earth and the Sun, casting shadows that create spectacular celestial events. A solar eclipse happens when the Moon passes between Earth and the Sun, temporarily blocking sunlight. A lunar eclipse occurs when the Moon passes into Earth's shadow, turning its surface a deep red hue. These events have inspired awe and wonder throughout history, providing opportunities for scientific observation and cultural reflection. The influence extends beyond the tides and eclipses; it plays a critical role in stabilizing Earth's axial tilt. Without the Moon, Earth's tilt could vary wildly over time, leading to dramatic shifts in climate. The stabilizing effect has helped create a relatively predictable environment, supporting the evolution of complex life.

One of the most puzzling aspects of the Moon is its magnetism. While the Moon lacks a global magnetic field today, ancient lunar rocks exhibit signs of magnetism, suggesting that the Moon once had a magnetic dynamo. The exact mechanism behind this ancient magnetic field remains a topic of ongoing research, with scientists exploring theories related to the core and its interactions with Earth's magnetic field.

The volcanic activity peaked billions of years ago, creating the maria that dominate its near side. These lava flows were driven by heat from radioactive decay in the interior. As the Moon cooled, volcanic activity ceased, leaving a geologically "dead" world. However, recent discoveries of small volcanic features suggest that the Moon may have experienced localized eruptions much more recently than previously thought. These findings challenge the view of the Moon as a completely inactive world and raise new questions about its thermal history.

The lack of a protective atmosphere allows it to interact directly with the solar wind, a stream of charged particles emitted by the Sun. These interactions have left traces on the lunar surface, including the formation of tiny bubbles in lunar rocks and the deposition of hydrogen ions, which may contribute to the formation of water molecules. The proximity and accessibility make it an ideal laboratory for studying planetary processes. Its surface preserves a record of the solar system's history, from the era of heavy bombardment to the quieter periods that followed. By studying the Moon, scientists can gain insights into the evolution of other rocky planets, including Earth, Mars, and Venus.

Geological History

The geological history is a story of fire and impact, a tale etched into its surface by ancient volcanic activity and relentless bombardment. From its fiery beginnings to its present state as a geologically "quiet" world, the Moon has undergone dramatic transformations that reveal the dynamic processes of planetary evolution. In the aftermath of the formation, its surface was dominated by a magma ocean—a vast, molten expanse resulting from the intense heat generated by the giant impact and the subsequent accumulation of material. As this magma ocean cooled, minerals began to crystallize and separate based on their densities.

The lighter minerals, such as plagioclase, floated to the surface, forming the initial crust. Heavier materials, including olivine and pyroxene, sank, creating the underlying mantle. This process of differentiation created the stratified structure that defines the interior today.

Approximately 4.1 to 3.8 billion years ago, the Moon experienced a period of intense asteroid and comet impacts, known as the Late Heavy Bombardment. This era was a chaotic time in the solar system, with planetary bodies colliding and exchanging material. On the Moon, these impacts created vast basins, including the Imbrium and Orientale basins, which remain some of its most prominent features.

The craters formed during this era provide a window into the solar system's early history. By studying these ancient scars, scientists can infer the frequency and intensity of impacts, as well as the origins of the objects that shaped the Moon.

Following the period of heavy bombardment, the Moon entered a phase of volcanic activity that lasted for hundreds of millions of years. Heat generated by radioactive decay in the mantle caused magma to rise to the surface, filling impact basins and creating the maria.

This volcanic activity was most prominent on the near side of the Moon, where the crust is thinner. The far side, with its thicker crust, experienced less volcanic activity, contributing to the stark differences in appearance between the two hemispheres.

The Surface: A Record of Time

The surface is a testament to its enduring relationship with the cosmos. Unlike Earth, where erosion and tectonic activity continually reshape the landscape, the surface is preserved in near-pristine condition, providing a record of events spanning billions of years.

One of the most valuable aspects of the surface is its craters, which serve as a kind of cosmic clock. By counting craters and analyzing their distribution, scientists can estimate the age of different regions. For example, the maria are relatively young, dating back about 3 to 4 billion years, while the highlands are much older, with some regions exceeding 4.5 billion years in age. The regolith, or lunar soil, is a fine layer of dust and small rock fragments that covers the surface. This material is formed by the constant bombardment of micrometeorites, which pulverize rocks and create a unique, glassy texture. The regolith is also enriched with tiny beads of volcanic glass, remnants of ancient eruptions that reveal the fiery past.

The chemical composition of the Moon provides crucial insights into its formation and evolution. Lunar rocks are dominated by silicate minerals, including feldspar and pyroxene, with trace amounts of metals and other elements.

One of the most intriguing discoveries from the Apollo missions was the presence of isotopic similarities between Earth and the

Moon. For example, the oxygen isotopic ratios of lunar and terrestrial rocks are nearly identical, supporting the idea that the Moon formed from material ejected by the giant impact. This isotopic "fingerprint" has become a cornerstone of our understanding of the Earth-Moon relationship.

The thermal evolution of the Moon is a complex process shaped by its size, composition, and proximity to Earth. Unlike Earth, which retains a molten core due to its larger size and active tectonics, the smaller size allowed it to cool more rapidly.

During its early history, the interior was hot enough to sustain volcanic activity, but as it cooled, its mantle became rigid and its volcanic eruptions ceased. Today, the Moon is considered geologically inactive, though recent studies suggest that it may still experience minor tectonic shifts, known as moonquakes, caused by the slow contraction of its cooling interior.

The influence on Earth extends beyond its gravitational pull. By stabilizing Earth's axial tilt, the Moon has helped maintain a relatively stable climate over geological timescales. This stability has been crucial for the development of complex ecosystems, as it has prevented extreme shifts in temperature and weather patterns.Moreover, the gravitational pull has shaped Earth's ocean currents, creating tidal patterns that influence marine ecosystems. These patterns play a critical role in distributing heat and nutrients across the globe, supporting biodiversity and regulating the planet's climate.

The Mystery of Lunar Water

One of the most exciting discoveries of recent decades is the presence of water on the Moon. Initially thought to be completely dry, the Moon has been revealed to harbor water in various forms, from ice in permanently shadowed craters to trace amounts bound in minerals. The discovery of water has profound implications for both science and exploration. Water is a critical resource for sustaining human life and could be used to produce oxygen and hydrogen for fuel. Understanding the origin and distribution of lunar water is a key focus of current and future missions, as it holds clues to the processes that delivered water to the inner solar system.

The unique characteristics make it an unparalleled laboratory for studying planetary processes. Its proximity to Earth allows for detailed observation and direct exploration, while its lack of atmosphere preserves a record of events that have shaped the solar system. For example, the craters provide insights into the history of asteroid impacts, helping scientists understand the potential threats posed by near-Earth objects. Similarly, the study of lunar rocks has revealed information about the processes of planetary differentiation and the early dynamics of the Earth-Moon system.

Despite decades of exploration, the Moon continues to pose questions that challenge our understanding of planetary science. For instance, the asymmetry between the near and far sides of the Moon remains a puzzle, as does the origin of its magnetic anomalies. These mysteries drive ongoing research and inspire new missions aimed at uncovering the secrets of our celestial companion. The story is inextricably linked to Earth's evolutionary timeline, serving as both a participant in and a recorder of critical planetary processes. Its formation marked the beginning of a gravitational relationship that would influence Earth's rotation, climate, and even the development of life.

When the Moon first formed, it orbited much closer to Earth, causing stronger tidal forces that significantly impacted the planet's rotation. Over time, the gravitational interaction between Earth and the Moon has slowed Earth's rotation, increasing the length of a day from just a few hours to the 24-hour cycle we know today. This slowing of Earth's rotation has implications for the planet's long-term stability. By distributing rotational energy to the Moon, causing it to drift farther away, the Earth-Moon system has evolved into a more stable configuration. This stability has contributed to Earth's ability to sustain complex ecosystems over billions of years.

The stabilizing effect on Earth's axial tilt has been vital for maintaining a relatively predictable climate. This is particularly important when considering the development of life, which thrives in stable environments. By preventing wild fluctuations in Earth's tilt, the Moon has helped ensure that seasonal changes remain moderate, creating conditions conducive to biodiversity. The lack of atmosphere and tectonic activity makes it a natural repository of ancient cosmic events. Unlike Earth, where geological processes erase evidence of impacts and other phenomena, the surface preserves a near-complete record of the solar system's history.

The craters provide a detailed timeline of asteroid and comet impacts. By studying these craters, scientists can infer the history of collisions in the inner solar system. This information is particularly valuable for understanding the Late Heavy Bombardment, a period when the inner planets were bombarded by debris left over from the formation of the solar system. The surface also retains evidence of its interaction with the solar wind and cosmic rays. Particles from the Sun and beyond leave traces in the regolith, creating a record of solar and galactic activity over billions of years. Analyzing these particles helps scientists understand changes in the Sun's behavior and the dynamics of the Milky Way.

The magnetic history is a subject of ongoing study, with implications for understanding its core and early evolution. While the Moon lacks a global magnetic field today, ancient lunar rocks exhibit signs of magnetism, suggesting that it once had a magnetic dynamo like Earth's. Certain regions of the Moon exhibit localized magnetic fields, known as magnetic anomalies. These anomalies, often associated with impact basins, may be remnants of an ancient magnetic field or the result of interactions with solar and cosmic phenomena. Understanding these anomalies could shed light on the thermal and geologic history.

The small, partially molten core contrasts with Earth's dynamic, fully molten core. This difference is a result of the smaller size, which allowed it to cool more rapidly. However, recent evidence suggests that the core may still generate minor heat, contributing to the weak tectonic activity observed today. The volcanic history is a testament to the dynamic processes that shaped its early evolution. While it is now considered geologically "dead," evidence of past volcanic activity can be seen in the maria, vast plains of basaltic rock that dominate the near side.

The maria were formed by massive lava flows that filled impact basins billions of years ago. These flows were driven by heat from radioactive decay in the mantle, which created reservoirs of magma that erupted onto the surface. The smooth, dark appearance of the maria contrasts sharply with the rugged, cratered highlands, reflecting the different geological histories of these regions.

In recent years, scientists have discovered small volcanic features on the Moon, suggesting that volcanic activity may have occurred more recently than previously thought. These features, such as small domes and pyroclastic deposits, challenge the idea that the Moon has been geologically inactive for billions of years.

The lack of an atmosphere exposes its surface to the full force of the solar wind, a stream of charged particles emitted by the Sun. This interaction has left a unique imprint on the surface, influencing its chemical composition and contributing to the formation of water molecules.

The solar wind implants hydrogen ions into the lunar regolith, which can combine with oxygen in minerals to form hydroxyl molecules. These molecules are a precursor to water and may account for some of the trace amounts of water detected on the surface. One of the most intriguing effects of the solar wind is the creation of lunar swirls—bright, curving patterns on the surface. These swirls are associated with magnetic anomalies that deflect the solar wind, protecting the underlying regolith from darkening. The study of lunar swirls provides insights into the magnetic history and its interaction with the solar environment.

The Relevance to Earth's Future

The Moon is not just a relic of the past; it holds significant potential for Earth's future. As humanity looks to establish a sustainable presence in space, the Moon offers a valuable proving ground for technologies and strategies that could support life on other worlds. The surface contains resources that could support future exploration and habitation. These include water ice in permanently shadowed craters, which could be used for drinking water, oxygen production, and fuel. Additionally, the regolith contains rare elements, such as helium-3, which has potential as a fuel for nuclear fusion. The proximity to Earth makes it an ideal location for scientific outposts. These outposts could serve as hubs for studying the Moon itself, as well as platforms for observing the cosmos. The absence of an atmosphere allows for clear, uninterrupted views of space, making the Moon an excellent location for telescopes and other observational instruments.

The Moon is also a stepping stone for deep space exploration. By establishing infrastructure on the Moon, humanity can develop the skills and technologies needed to venture to Mars and beyond. The low gravity makes it an efficient launch point for missions to other planets, reducing the energy and resources required for interplanetary travel. One of the most striking features is the stark difference between its near and far sides. This asymmetry, first revealed by spacecraft missions in the mid-20th century, has puzzled scientists for decades and provides key insights into the formation and evolution.

The side of the Moon visible from Earth, known as the near side, is dominated by dark maria and bright highlands. The maria cover about 31% of the near side, forming vast, basaltic plains that were created by ancient volcanic activity. These smooth regions contrast sharply with the rugged terrain of the highlands, which are heavily cratered and composed of lighter-colored anorthosite. In contrast, the far side of the Moon is almost entirely covered by highlands, with maria accounting for less than 2% of its surface. This difference is thought to be the result of variations in crustal thickness between the two hemispheres. The far side's crust is significantly thicker, likely due to asymmetries in the early formation, which may have inhibited volcanic activity in this region.

Several hypotheses have been proposed to explain the asymmetry. One leading theory suggests that the impact that formed the Moon distributed heat and material unevenly, causing the near side to cool more slowly and retain more radioactive elements. This "Procellarum KREEP Terrane" on the near side contains higher concentrations of potassium, rare earth elements, and phosphorus, which may have fueled prolonged volcanic activity. Another theory posits that tidal interactions with Earth played a role in shaping the asymmetry. The near side's proximity to Earth may have caused additional heating, influencing its geological development. The gravitational and physical characteristics have had profound effects on Earth, shaping its geological and biological history. The gravitational pull generates tides, which play a critical role in Earth's climate and ecosystems. Tides influence ocean currents, which help regulate global temperatures and distribute nutrients. These processes have been vital for sustaining marine life and supporting the biodiversity that thrives in tidal zones.

Tides may have also influenced the early development of life on Earth. The cyclical nature of tidal movements created dynamic environments in coastal areas, exposing organisms to varying conditions. These fluctuations could have driven evolutionary adaptation, encouraging the development of traits such as mobility and resilience. Earth's axial tilt, or obliquity, is stabilized by the gravitational pull. This stabilization has prevented dramatic shifts in Earth's orientation, ensuring a relatively predictable pattern of seasons. Without the Moon, Earth's tilt could vary wildly over time, leading to extreme climatic changes that would make the planet less hospitable for life.

The influence on Earth's tilt is not constant; it gradually decreases as the Moon moves farther away. Over billions of years, this slow drift will alter the dynamics of the Earth-Moon system, potentially impacting Earth's long-term climate stability.

The lunar regolith, the fine layer of dust and rock fragments covering the surface, is a treasure trove of scientific information. Formed by billions of years of micrometeorite impacts and space weathering, the regolith preserves a record of the history and its interaction with the solar environment.

The constant bombardment of micrometeorites has pulverized the surface, creating a layer of regolith that varies in thickness across different regions. This process has also produced tiny glass beads, known as agglutinates, which provide clues about the conditions and composition of the lunar surface.

The exposure to the solar wind has implanted elements such as hydrogen, helium, and neon into the regolith. These elements, carried by the Sun's energetic particles, are trapped in the lunar soil and offer insights into the Sun's activity over time. For instance, studying the isotopic composition of helium in the regolith can reveal patterns of solar wind intensity and variability.

While the Moon lacks active plate tectonics, it has experienced significant volcanic and tectonic activity in its past. These processes have left behind features that provide a window into the thermal and geological evolution.

The vast lava plains of the maria are the most prominent volcanic features on the Moon. These plains were formed by massive eruptions that occurred between 3.8 and 1 billion years ago. The maria are rich in basalt, a dark volcanic rock that contrasts with the lighter anorthosite of the highlands.

Other volcanic features include sinuous rilles, which are long, winding channels carved by flowing lava. These rilles, such as the famous Schröter's Valley, provide evidence of the once-dynamic interior.

The surface also exhibits tectonic features, such as wrinkle ridges and graben, which formed as the crust responded to internal stresses. Wrinkle ridges are the result of compressional forces that created folds in the crust, while graben are formed by extensional forces that caused the crust to fracture and sink.

The Magnetic Anomalies

One of the most enigmatic aspects of the Moon is the presence of localized magnetic fields, known as magnetic anomalies. These anomalies are concentrated in specific regions, such as the Reiner Gamma formation, and are thought to be remnants of an ancient magnetic field.

The early molten core may have generated a magnetic field through a dynamo process, similar to Earth's. As the core cooled and solidified, this dynamo activity ceased, leaving behind localized regions of magnetized rock. Understanding these anomalies is critical for reconstructing the thermal and magnetic history.

Magnetic anomalies could also have practical implications for future lunar exploration. These regions may provide partial protection from cosmic radiation, making them potential sites for human habitation or scientific outposts.

As the closest celestial body to Earth, the Moon serves as a gateway to understanding planetary processes throughout the solar system. Its preserved surface and proximity allow for detailed observation and exploration, making it an invaluable resource for planetary science.

The geological features, such as craters and lava plains, provide a baseline for studying similar processes on other planets and moons. By comparing the history to that of Mars, Venus, and Mercury, scientists can gain insights into the diversity of planetary evolution. The Moon also serves as a testing ground for technologies that will be critical for exploring more distant destinations, such as Mars. From robotic landers to human habitats, the lessons learned on the Moon will inform the next generation of space exploration.

Chapter 3: The Moon Through History

From the dawn of civilization, the Moon has been a source of wonder and inspiration. Its silvery glow, changing phases, and mysterious eclipses captured the imaginations of ancient peoples, who wove the Moon into their myths, calendars, and rituals. Long before telescopes and scientific instruments, the Moon was humanity's first guide to understanding the cosmos.

In many cultures, the Moon was seen as a divine entity or a powerful force shaping the world. The ancient Greeks personified the Moon as Selene, a goddess who traversed the night sky in a silver chariot. In Chinese mythology, the Moon is home to Chang'e, a goddess who fled there after consuming an elixir of immortality. Native American tribes often associated the Moon with cycles of life and renewal, reflecting its influence on nature and time.

The phases symbolized change and continuity, appearing in art, literature, and spiritual practices. Full moons were celebrated in festivals, while new moons marked beginnings. Across the globe, the Moon became a unifying symbol of mystery and connection, linking diverse cultures through its universal presence in the night sky.

Early humans relied on the Moon to measure time, using its predictable phases to create calendars. Lunar calendars, based on the 29.5-day cycle from new moon to new moon, were among the first systems of timekeeping. These calendars guided agricultural practices, religious observances, and daily life.

For example, the Islamic calendar remains a purely lunar calendar, determining the timing of Ramadan and other important events. Similarly, the Chinese calendar combines lunar and solar elements, linking traditional festivals to the phases.

As human understanding of the natural world evolved, the Moon became a focal point of early scientific inquiry. Ancient astronomers observed the movements, recognizing its connection to tides and its role in eclipses. These observations laid the groundwork for the development of astronomy.

The Babylonians were among the first to document the motions with precision. Using detailed records of lunar phases and eclipses, they developed mathematical models to predict celestial events. Their work influenced later civilizations, including the Greeks and Romans, shaping the foundations of Western astronomy.

In ancient Greece, philosophers and scientists sought to explain the nature and its relationship to Earth. Pythagoras proposed that the Moon reflected sunlight, while Aristotle argued that the Moon was a spherical body, based on its curved shadow during lunar eclipses. The Roman author Pliny the Elder expanded on these ideas in his *Natural History*, describing the phases and its effect on tides. These early theories, though limited by observational technology, demonstrated a growing curiosity about the role in the cosmos.

The Renaissance marked a turning point in humanity's understanding of the Moon. With the invention of the telescope, astronomers could observe the surface in unprecedented detail, challenging long-held beliefs and sparking a revolution in science. In 1609, Galileo Galilei turned his telescope to the Moon, becoming the first person to document its surface features. His sketches revealed mountains, craters, and valleys, proving that the Moon was not a smooth, divine sphere but a rugged, physical world.

Galileo's observations were revolutionary, challenging the prevailing view of a perfect celestial realm. His work inspired other scientists to study the Moon, laying the groundwork for modern lunar science. Johannes Kepler, a contemporary of Galileo, studied the mechanics of lunar eclipses, using them to refine his theories of planetary motion. Kepler's work demonstrated the interconnectedness of celestial phenomena, linking the behavior to the broader dynamics of the solar system.

As scientific knowledge advanced, so did humanity's ability to explore and understand the Moon. The 19th and early 20th centuries saw significant breakthroughs in lunar science, driven by advances in technology and a growing interest in space exploration. With the invention of more powerful telescopes, astronomers began creating detailed maps of the surface. In 1834, Johann Heinrich Mädler and Wilhelm Beer published the first accurate lunar atlas, which included precise measurements of lunar features. Their work provided a foundation for future exploration, allowing scientists to identify potential landing sites and study the geology. The advent of photography revolutionized lunar observation. In 1840, John William Draper captured the first photograph of the Moon, opening a new era of visual documentation. By the late 19th century, astronomers were using photography to create detailed images of the craters, mountains, and plains.

The mid-20th century marked the beginning of the Space Age, transforming humanity's relationship with the Moon. What was once a distant object of curiosity became a tangible destination, as nations raced to explore and conquer the lunar frontier.

The Space Race between the United States and the Soviet Union was fueled by Cold War rivalries and a desire to demonstrate technological superiority. In 1959, the Soviet Luna 2 mission became the first human-made object to impact the Moon, followed by Luna 3, which captured the first images of the far side. The United States responded with the Apollo program, culminating in the historic Apollo 11 mission. On July 20, 1969, Neil Armstrong and Buzz Aldrin became the first humans to set foot on the Moon, marking a triumph of human ingenuity and exploration. The Apollo missions provided an unprecedented wealth of data, including lunar rocks and soil samples that revealed the geological history. These findings confirmed the giant impact hypothesis, deepened our understanding of planetary formation, and highlighted the importance in Earth's evolution.

The influence extends beyond science and exploration; it remains a powerful symbol in art, literature, and popular culture. From the poetry of Emily Dickinson to the music of Pink Floyd, the Moon has inspired countless creative works, reflecting its enduring presence in human imagination.

The cycles and phases have long been a metaphor for change, growth, and renewal. Writers, musicians, and artists have drawn on its symbolism, using the Moon to explore themes of love, longing, and mystery.

In modern times, the Moon has become a popular setting for science fiction, representing the first step toward humanity's interplanetary future. Films like *2001: A Space Odyssey* and novels like Arthur C. Clarke's *Earthlight* envision the Moon as a frontier of discovery and adventure.

The Moon has always been more than a celestial body; it is a mirror of human emotion, a symbol woven into the fabric of stories, traditions, and dreams. Across cultures and epochs, its pale glow has served as both a guide and a muse, inspiring humanity to look outward and inward.

The Moon featured prominently in the spiritual and cultural practices of ancient civilizations. In Mesopotamia, the Moon was personified as the god Nanna (later called Sin), who represented wisdom and fertility. Temples dedicated to Nanna were centers of learning and observation, emphasizing the perceived power over human and agricultural cycles.

Similarly, in ancient Egypt, the Moon was associated with the god Thoth, who governed wisdom, time, and the arts. Lunar cycles influenced religious ceremonies, agricultural planting, and even medical practices, reflecting the integration into daily life.

In the Americas, the Moon held equal reverence. The Inca civilization saw the Moon as a goddess, Mama Quilla, the wife of the Sun god Inti. She was believed to govern time, marriage, and fertility, her phases marking the rhythm of life in the Andes.

The Moon in Early Exploration and Navigation

Before the advent of modern instruments, the Moon played a crucial role in navigation and timekeeping. Mariners relied on the phases to track tides, while its steady progression through the night sky served as a celestial clock.

The regular cycles offered a dependable measure of time, particularly for cultures without advanced astronomical tools. Ancient Polynesians, for example, used the Moon to navigate vast oceanic distances, relying on its phases to plan their voyages and align their journeys with seasonal changes.

In ancient China, the Moon was central to the lunar zodiac, a system that combined lunar cycles with the positions of celestial bodies. This zodiac guided agricultural practices, festivals, and personal fortunes, emphasizing the connection to human activity. As human societies advanced, the Moon became a focal point for scientific inquiry. Its motions, phases, and eclipses provided opportunities for observation, measurement, and theoretical exploration.

One of the earliest scientific breakthroughs involving the Moon was the understanding of lunar eclipses. Ancient Greek astronomers, including Anaxagoras and Hipparchus, recognized that lunar eclipses occurred when the Moon passed through Earth's shadow. This realization led to early attempts to measure the relative sizes and distances of Earth, the Moon, and the Sun.

The Greek astronomer Aristarchus of Samos used the Moon to propose a heliocentric model of the solar system as early as the 3rd century BCE. By observing the phases and calculating its distance from Earth, Aristarchus suggested that the Sun, not Earth, was the center of the universe—a revolutionary idea that was largely dismissed until the Renaissance.

The Renaissance and the Birth of Lunar Science

The Renaissance era reignited humanity's fascination with the Moon, combining artistic inspiration with scientific innovation. The invention of the telescope transformed the Moon from a distant mystery into a tangible world, laying the foundation for modern astronomy.

In 1610, Galileo Galilei published *Sidereus Nuncius* (*Starry Messenger*), a groundbreaking work that revealed the rugged terrain. Using his telescope, Galileo identified mountains, valleys, and craters, challenging the Aristotelian view of celestial perfection. Galileo's observations had profound implications, not only for lunar science but also for humanity's perception of the cosmos. The Moon was no longer an ethereal realm but a physical place, subject to the same natural laws as Earth.

Johannes Kepler, a contemporary of Galileo, expanded on the understanding of the motion. Kepler's laws of planetary motion, derived from meticulous observations, explained the elliptical orbit and its gravitational relationship with Earth. These laws laid the groundwork for future studies of orbital mechanics and space exploration.

The Moon as a Scientific Frontier

The Enlightenment brought a new era of systematic inquiry, with the Moon serving as a key focus of observation and experimentation. Advances in optics, mathematics, and physics enabled scientists to study the Moon with greater precision. Isaac Newton's *Principia Mathematica*, published in 1687, revolutionized the study of the Moon by introducing the concept of universal gravitation. Newton demonstrated that the orbit around Earth was governed by the same force that caused objects to fall on Earth—gravity. This insight unified celestial and terrestrial mechanics, providing a framework for understanding the motion of the Moon and other celestial bodies.

The 17th and 18th centuries saw the rise of lunar cartography, as astronomers created detailed maps of the surface. Notable figures, such as Giovanni Battista Riccioli and Johannes Hevelius, documented the features, naming its craters, seas, and mountains. These maps not only advanced scientific knowledge but also fueled the imagination of writers and explorers who dreamed of visiting the Moon.

The Industrial Revolution ushered in a new era of technological advancement, enabling more precise observations of the Moon. Larger telescopes, improved lenses, and photographic techniques allowed astronomers to capture detailed images of the lunar surface. In 1839, Louis Daguerre's invention of the daguerreotype made it possible to photograph the Moon. This innovation allowed astronomers to document its features with unprecedented accuracy, creating a visual archive that could be studied and compared over time.

Prominent observatories, such as the Royal Observatory in Greenwich and the Yerkes Observatory in the United States, focused their efforts on the Moon. Astronomers like William Herschel and Edward Emerson Barnard used advanced telescopes to study lunar craters, maria, and other features, contributing to a deeper understanding of the geology.

By the late 19th century, the Moon had become a symbol of humanity's desire to explore beyond Earth. Writers such as Jules Verne and H.G. Wells envisioned journeys to the Moon, blending scientific speculation with imaginative storytelling. These works reflected the growing belief that lunar exploration was not only possible but inevitable.

Jules Verne's *From the Earth to the Moon* (1865) and H.G. Wells's *The First Men in the Moon* (1901) captivated readers with their visions of lunar voyages. These stories, though fictional, inspired generations of scientists, engineers, and explorers, bridging the gap between fantasy and reality.

By the turn of the 20th century, the development of rocketry and advancements in physics brought humanity closer to realizing the dream of reaching the Moon. The groundwork was laid for the Space Age, a period that would transform the Moon from a distant observer into an active participant in human history.

Chapter 4: Moon's Role in the Solar System

The existence has shaped Earth's development in profound ways. From its gravitational influence to its impact on Earth's stability and habitability, the Moon is a cornerstone of our planet's evolutionary journey.

One of the most critical roles is stabilizing Earth's axial tilt. Without the Moon, Earth's tilt could vary wildly over time, resulting in chaotic climate shifts. The gravitational pull acts as a stabilizing force, keeping Earth's axial tilt relatively consistent at 23.5 degrees. This stability has been essential for maintaining predictable seasons and a hospitable climate for billions of years.

The gravitational interaction between Earth and the Moon is not static; it evolves over time. Tidal forces generated by the Moon gradually transfer energy from Earth's rotation to the orbit. This process has slowed Earth's rotation, increasing the length of a day from approximately 6 hours in the planet's early history to the current 24-hour cycle. Simultaneously, the Moon is moving away from Earth at a rate of about 3.8 centimeters per year.

The gravitational pull is the primary driver of Earth's ocean tides. By creating periodic rises and falls in sea levels, the Moon has influenced coastal ecosystems, sediment distribution, and even the evolution of marine life. The interplay of lunar and solar gravitational forces also creates spring tides and neap tides, adding complexity to these natural rhythms.

The Moon is not just important for Earth—it serves as a model for understanding planetary evolution across the solar system. The origin through a giant impact provides insights into the early chaotic phase of planetary formation. This hypothesis suggests that collisions between protoplanets and planetesimals were common in the young solar system, shaping the composition and dynamics of planets and their satellites.

By studying the Earth-Moon system, scientists have developed models to explain the formation of other moons, such as those of Mars, Jupiter, and Saturn. The lessons learned from the origin offer a framework for understanding the diversity of moons and planetary systems beyond our own.

Unlike Earth, where erosion and tectonics erase geological records, the surface preserves a nearly pristine history of the solar system. Its craters and basins provide a timeline of asteroid impacts, while its regolith contains evidence of solar wind interactions and cosmic radiation.

By studying the Moon, scientists can reconstruct the early solar system's bombardment history and gain insights into the conditions that shaped planets like Earth.

The gravitational and tidal effects have played a role in shaping Earth's biosphere, influencing the development and behavior of life. The tidal forces create dynamic environments in coastal areas, where tidal pools experience regular cycles of flooding and exposure. These environments are thought to have provided fertile ground for the chemical reactions that led to the origin of life. The cyclical nature of tides may have concentrated organic molecules, promoting the formation of complex compounds and primitive life forms.

The phases and cycles have been ingrained in the behavior of many species. For example, some marine organisms, such as corals and certain fish, synchronize their reproductive activities with the lunar cycle. These adaptations reflect the deep evolutionary influence of the Moon on Earth's biosphere. The gravitational relationship with Earth and the Sun creates complex dynamics that affect the entire Earth-Moon-Sun system.

The alignment of the Moon, Earth, and the Sun produces spectacular phenomena such as solar and lunar eclipses. These events are not only awe-inspiring but also provide opportunities to study the mechanics of celestial motion. The precise alignment required for eclipses reveals the intricate balance of gravitational forces within the solar system.

The orbit is not a perfect circle but an ellipse, causing its distance from Earth to vary slightly over time. This variation, combined with the tilt of its orbit, produces a phenomenon called libration, where the Moon appears to "wobble" as seen from Earth. Libration allows observers to glimpse slightly more than half of the surface, revealing additional details for study.

The position and behavior within the solar system make it an important player in understanding broader planetary interactions. The Earth-Moon system is one of many examples of planet-satellite relationships in the solar system. By comparing the Moon to other natural satellites, such as Jupiter's volcanic Io or Saturn's icy Enceladus, scientists can identify commonalities and differences in their formation, evolution, and potential for hosting life.

The gravitational pull affects not only Earth but also objects in its vicinity, such as asteroids and comets. While the role as a shield against asteroid impacts is limited compared to Jupiter, its presence still influences the trajectories of near-Earth objects. Understanding these interactions is critical for assessing the potential threat of asteroid collisions.

The proximity to Earth and its unique characteristics make it an ideal destination for exploration and a stepping stone for further space missions. The Moon's surface contains valuable resources, such as water ice, which could support long-term human presence in space. By extracting and utilizing these resources, future missions could become more self-sufficient, reducing the need for resupply from Earth.

Establishing scientific outposts on the Moon would enable continuous observation of its surface and the broader cosmos. These outposts could serve as hubs for studying planetary science, testing life-support systems, and training astronauts for future interplanetary missions.

The relationship between Earth and its satellite has shaped not only the planet but the broader celestial mechanics of our solar system. This interaction is a delicate dance of gravity, motion, and time that has profound implications beyond the immediate partnership of these two celestial bodies. By influencing axial stability, tides, and the distribution of mass across the globe, this gravitational relationship acts as a force that maintains balance within a dynamic system.

Without the constant gravitational stabilizer, Earth's axis could shift erratically, creating extreme seasonal changes over millennia. These shifts would have profound consequences for global climates, likely disrupting the evolutionary processes that relied on relative predictability. The relationship between these two celestial bodies ensures the steadiness needed for ecosystems to adapt and flourish over long epochs.

The gravitational forces extend far beyond the immediate relationship, affecting nearby celestial objects and even the trajectories of asteroids and comets. These minor adjustments to trajectories demonstrate how interconnected bodies within the solar system subtly influence one another. The ripple effects of these gravitational nudges contribute to the overall dynamic balance that governs planetary orbits and their satellites.

The forces at play are not static; energy transfers between celestial partners result in gradual changes that unfold over eons. The rotation of Earth, for example, continues to slow as tidal friction dissipates energy into the system. This energy loss is offset by the slow outward drift of its satellite, a phenomenon measurable with modern laser-ranging techniques. While small in the short term, these incremental changes accumulate over billions of years, shaping the current equilibrium.

The wider solar system provides parallels and contrasts that further illuminate the unique characteristics of this gravitational relationship. For instance, comparisons with the many moons of Jupiter and Saturn reveal the importance of size, composition, and distance in shaping the evolution of these systems. The intense gravitational interplay among the gas giants and their many satellites creates phenomena not seen elsewhere, such as tidal heating and active volcanism. By studying these systems, scientists gain insights into what makes Earth's satellite and its influence on its host so distinct.

Exploration efforts have brought humanity closer to understanding the larger dynamics at play. Scientific missions, from early robotic landers to sophisticated orbiters, have revealed a wealth of data about the satellite's geological history and its role in the broader cosmos. These missions are more than exercises in curiosity—they are windows into the processes that govern planetary systems across the galaxy.

The same forces shaping this celestial duo also guide efforts to explore the possibilities of long-term human settlement and utilization. Understanding the fine balance of gravitational and orbital mechanics is essential for constructing sustainable infrastructure in space. From resource extraction to low-gravity launching points for interplanetary missions, future developments will hinge on the precise understanding of these dynamics.

In the broader context of the solar system, the satellite's role as a recorder of cosmic history is unparalleled. The surface, largely unchanged by the erosion and tectonic activity that erase evidence on Earth, acts as a time capsule. Impact craters, ancient lava flows, and even the faint remnants of solar radiation are preserved, offering an unbroken timeline of events that shaped the inner planets. By examining this natural archive, scientists can infer details about the period of heavy bombardment that affected all terrestrial planets.

This record is not simply a scientific curiosity. Understanding the frequency and intensity of early impacts offers practical insights into planetary defense strategies. Near-Earth objects, whose trajectories are influenced by gravitational interplay, remain a persistent concern for modern scientists. Studying historical patterns in the system provides a clearer picture of the risks posed by these wandering celestial bodies.

Looking ahead, the role of Earth's satellite extends into the future of human exploration. Its low-gravity environment makes it an ideal launching point for interplanetary missions, requiring less energy to escape its gravitational pull. This strategic advantage positions it as a central hub in humanity's broader ambitions in space, from building sustainable bases to fueling missions aimed at Mars and beyond.

As humanity reaches further into the solar system, the lessons learned from this dynamic relationship will continue to inform the designs and strategies of exploration. The satellite will remain both a partner in understanding the past and a guide to navigating the future, shaping the trajectory of discovery across the solar system and beyond.

The connection between Earth and its natural satellite extends beyond immediate gravitational effects, deeply embedding itself into the intricate mechanics of the solar system. This relationship, formed during the chaotic early stages of planetary development, has contributed to shaping not only Earth's present conditions but also its long-term evolution. These effects, subtle yet profound, are observable across geological, climatic, and even cosmic timescales.

Gravitational influences are the most apparent aspect of this relationship. By exerting steady force over billions of years, the satellite contributes to the regulation of axial tilt, ensuring that Earth avoids extreme seasonal variability. This balance has created a stable environment conducive to life's persistence and evolution. Without this moderating effect, dramatic shifts in axial tilt could lead to long periods of glaciation or searing heat, disrupting the delicate equilibrium that sustains ecosystems.

Tidal forces, another significant contribution, extend their influence beyond merely raising and lowering ocean levels. The periodic movement of tides affects sediment distribution, coastal erosion, and nutrient mixing, all of which are essential to maintaining marine biodiversity. These forces also contribute to the creation of unique habitats, such as tidal marshes, which play crucial roles in supporting species diversity and acting as buffers against environmental changes.

This connection between celestial partners is not isolated; it reverberates throughout the solar system. The combined gravitational pull from Earth and its satellite subtly alters the trajectories of passing asteroids and comets. This interaction occasionally deflects potential impacts or, conversely, draws objects closer, shaping the near-Earth environment over time. Such gravitational influences, though often unnoticed, underline the broader interplay of forces that govern planetary systems.

Comparisons with other celestial systems further emphasize the uniqueness of this relationship. The dynamic interactions among the moons of Jupiter, for instance, demonstrate the intense gravitational tug-of-war created by massive host planets. While these systems exhibit phenomena like tidal heating and subsurface oceans, Earth's companion stands apart in its stabilizing influence and its role in fostering a planet's long-term habitability. These comparisons are essential in understanding the conditions necessary for life elsewhere in the universe.

The lack of atmospheric and tectonic activity on the surface allows it to act as a cosmic time capsule, preserving the history of the inner solar system. Impact craters spanning billions of years provide a snapshot of the era of heavy bombardment, a time when the inner planets were subjected to intense collisions that shaped their surfaces and compositions. Studying these craters offers insights not only into Earth's early history but also into the broader processes that govern planetary formation and evolution.

The geochemical record preserved in the regolith—the thin, powdery layer covering the surface—reveals interactions with solar wind and cosmic radiation over time. By analyzing isotopic compositions and trapped particles, scientists can reconstruct changes in solar activity and radiation levels, data crucial for understanding the Sun's behavior and its impact on the inner planets.

The surface's role as a repository of cosmic history extends to its potential for future exploration and utilization. The discovery of water ice in shadowed regions has transformed perceptions of its potential for supporting human activity. This resource, vital for both sustenance and fuel production, represents a cornerstone for developing sustainable exploration strategies. By utilizing locally sourced materials, future missions can reduce dependence on Earth-based supplies, paving the way for long-term settlement and deeper space ventures.

Humanity's exploratory ambitions are further supported by the low gravity, which makes launching spacecraft significantly more efficient. This advantage, combined with its proximity to Earth, positions the satellite as an ideal staging ground for missions to more distant destinations. The infrastructure established here could serve as a blueprint for developing similar systems on Mars or other celestial bodies, creating a foundation for a broader human presence in the solar system.

In addition to its practical applications, the exploration of this celestial body continues to inspire scientific inquiry and innovation. Missions that investigate its surface, interior, and gravitational interactions yield insights that extend far beyond the immediate system, informing our understanding of planetary systems across the galaxy. As technology advances, these studies will deepen, uncovering new facets of its role in the cosmic order. Looking ahead, the partnership between Earth and its companion will remain a focal point for understanding the forces that shape planetary systems. Its role as both a stabilizer and a witness to cosmic history ensures its continued significance, not only in the context of Earth's past and present but also in humanity's future as explorers of the universe.

The gravitational relationship between Earth and its satellite has shaped the behavior of the solar system in ways both subtle and profound. This influence extends beyond tides and axial stability, impacting planetary orbits, asteroid trajectories, and even the evolution of life. These interactions are reminders of the interconnectedness of celestial mechanics, where no body exists in isolation.

The satellite's role in slowing Earth's rotation over billions of years is a prime example of this interdependence. Tidal friction, caused by gravitational interactions, transfers rotational energy, leading to longer days and a more distant orbit. This energy exchange has been critical to the evolution of Earth's ecosystems. Slower rotation has moderated wind patterns, influenced ocean currents, and created conditions favorable for complex life. These changes, while imperceptible in human timescales, underscore the enduring impact of gravitational forces.

Within the broader solar system, the presence of a large companion contributes to Earth's gravitational footprint. Together, they influence the paths of nearby asteroids and comets, creating a gravitational buffer zone that can alter the trajectory of potentially hazardous objects. This protective effect, while not absolute, has likely mitigated the frequency of impacts, preserving Earth's surface for life's continuity.

The satellite's surface, marked by craters and plains, offers a preserved record of these interactions. Unlike Earth, where erosion and tectonics erase evidence of impacts, this celestial body retains a pristine history of bombardment. These scars provide insight into the intensity of asteroid activity during the early solar system, a period critical to understanding planetary formation.

The preserved layers of regolith hold another layer of cosmic history. Particles from the Sun, trapped in this material, create a timeline of solar wind activity and radiation levels. By analyzing these records, scientists gain a clearer picture of the Sun's behavior over millennia, data essential for understanding the conditions that shaped the inner planets.

In the context of planetary science, the satellite serves as a comparative model. Its lack of active geology allows scientists to study features formed billions of years ago, offering a baseline against which to compare other planetary bodies. This comparative approach has illuminated processes such as volcanic activity, crustal formation, and impact dynamics, enriching our understanding of rocky worlds.

As humanity advances its exploration capabilities, this celestial companion continues to offer unique advantages. Its proximity and low gravity make it an ideal testing ground for technologies and strategies needed for deep-space exploration. For example, systems for extracting water from ice deposits or using regolith as construction material are being developed with long-term habitation in mind. These innovations could one day support human missions to Mars or even further afield.

Its role as a base for scientific observation is equally significant. The absence of an atmosphere allows for clear, uninterrupted views of the cosmos, making it a prime location for telescopes and other instruments. Observatories established here could revolutionize our understanding of the universe, from the behavior of distant galaxies to the detection of exoplanets.

In addition to its scientific and practical benefits, this celestial body represents humanity's first step toward becoming an interplanetary species. The knowledge gained from studying its surface, using its resources, and understanding its place in the solar system will guide our efforts to explore and inhabit other worlds.

While its historical role has been largely one of stability and observation, its future is intertwined with humanity's aspirations. As we seek to expand beyond our home planet, this ancient partner will remain at the forefront of discovery, offering both a gateway and a guide to the vast unknowns of space.

Chapter 5: Exploring the Moon

Humanity's journey to explore its celestial neighbor began long before the advent of rockets and space agencies. For centuries, the surface features visible from Earth have sparked curiosity and speculation. From ancient astronomers sketching its craters to robotic missions mapping its terrain, the process of exploration has been a story of progress, ambition, and perseverance. The exploration of this body has not only expanded our understanding of the cosmos but also redefined what humanity can achieve.

The modern era of exploration began with the Space Age in the mid-20th century. As Cold War tensions drove technological competition, space exploration became a symbol of national pride and scientific achievement. Early robotic missions provided the first close-up views, paving the way for human landings. Each mission, whether robotic or manned, has contributed to an evolving picture of this celestial body's geology, history, and potential.

The Soviet Union's Luna program marked the first major milestone in modern exploration. In 1959, Luna 2 became the first human-made object to impact its surface, and Luna 3 captured the first images of the far side. These early missions revealed a rugged, cratered landscape and confirmed many long-standing theories about its nature. These achievements sparked global fascination, igniting dreams of sending humans beyond Earth.

The Apollo program represented a turning point in exploration. Between 1969 and 1972, six manned missions successfully landed on the surface, culminating in unprecedented scientific discoveries. Apollo 11's historic landing in 1969, when Neil Armstrong and Buzz Aldrin stepped onto the surface, fulfilled a goal that had captivated humanity for generations. The samples returned from these missions revealed a complex geological history, confirming the hypothesis that the satellite shared a common origin with Earth.

While Apollo captured the world's attention, subsequent missions continued to push the boundaries of robotic exploration. NASA's Lunar Reconnaissance Orbiter, launched in 2009, provided high-resolution maps of its surface, identifying potential landing sites for future missions. Other programs, such as India's Chandrayaan and China's Chang'e series, have added valuable data about its composition and resource potential.

One of the most significant discoveries of recent decades has been the presence of water ice in permanently shadowed regions near the poles. This finding has profound implications for future exploration, as water is a vital resource for sustaining human life and producing fuel. The discovery has shifted the focus of many exploration programs toward these polar regions, which were previously considered inhospitable due to their extreme cold.

The technological advancements driving exploration have been matched by evolving goals. Early missions focused on gathering basic scientific data, such as surface composition and geological features. Today, the emphasis is on sustainability and preparation for long-term human presence. Concepts such as in-situ resource utilization—using local materials for building and sustaining life—are central to these efforts.

International cooperation has become a hallmark of modern exploration. The Artemis program, led by NASA with contributions from multiple countries, aims to establish a sustainable human presence. By combining resources and expertise, these collaborative efforts promise to accelerate progress and ensure that exploration benefits all of humanity.

Looking ahead, the exploration of this celestial body serves as a precursor to more ambitious goals. Its proximity and relatively simple environment make it an ideal testing ground for technologies and strategies needed for missions to Mars and beyond. Lessons learned from its exploration will inform every aspect of interplanetary travel, from habitat construction to life support systems.

Yet exploration is not merely a technical endeavor. It is a reflection of humanity's inherent desire to understand and push boundaries. Every step taken on this celestial surface represents a leap forward in our quest to comprehend the universe and our place within it. As exploration continues, it will undoubtedly uncover new mysteries and possibilities, inspiring future generations to dream even bigger.

The exploration of our celestial neighbor represents a journey into the unknown, fueled by human curiosity and the relentless pursuit of knowledge. The strides taken in the mid-20th century during the Space Race set the stage for a new age of discovery, with each mission building upon the successes and lessons of its predecessors.

The Soviet Union's early accomplishments through the Luna program provided critical insights and paved the way for more complex missions. Luna 9, in 1966, was the first spacecraft to achieve a soft landing, sending back images of the surface and proving that it could support the weight of future landers. This mission confirmed key hypotheses about the surface's composition and texture, dispelling fears of landing in dangerously soft regolith.

The United States' Apollo program remains one of the most iconic achievements in human history. Beyond the historic first steps of Apollo 11, the program contributed vast scientific knowledge. Each mission, from Apollo 12 to Apollo 17, carried advanced scientific equipment and returned with samples that transformed our understanding of the satellite's geological history. These missions provided evidence supporting the giant impact hypothesis, shedding light on the shared origins of Earth and its companion.

One of the lesser-known yet critical aspects of the Apollo missions was the deployment of instruments for long-term observation. These included seismometers, which detected "moonquakes" and revealed that its interior was far more complex than previously believed. These data suggested a partially molten core and active thermal processes, challenging the notion of the satellite as a completely inert body.

After the Apollo era, exploration entered a quieter phase, marked primarily by robotic missions. However, these missions were no less significant. The Clementine mission in the 1990s reignited interest in lunar exploration by mapping the surface with unprecedented detail and detecting hints of water ice near the poles. This discovery set the stage for a renewed focus on polar regions, where permanently shadowed craters might harbor resources critical for future human activity.

In recent decades, international missions have expanded humanity's reach and understanding. India's Chandrayaan-1, launched in 2008, confirmed the presence of water molecules on the surface, a groundbreaking finding that reshaped exploration strategies. China's Chang'e program has achieved remarkable milestones, including the first landing on the far side and the return of lunar samples in 2020. These achievements highlight the growing capabilities and ambitions of emerging space powers.

The Artemis program signals a bold new chapter in exploration, aiming to establish a sustainable human presence. With plans to land the first woman and the next man on the surface, Artemis represents a collaborative effort involving multiple nations and private companies. The program's long-term vision includes building infrastructure for extended stays, such as habitats and power systems, which will serve as a blueprint for interplanetary missions.

Technological advancements are central to this new era. Autonomous rovers, precision landers, and advanced propulsion systems have significantly expanded the scope of what is possible. Concepts such as 3D printing using regolith are being developed to construct habitats and tools on-site, reducing the need for costly launches from Earth. These innovations are not only advancing science but also addressing the practical challenges of living and working on a distant body.

The discovery of water ice has been a game changer, transforming exploration from a short-term endeavor to a long-term vision. Water is essential not only for human survival but also for producing oxygen and hydrogen, which can be used as rocket fuel. This potential to "live off the land" marks a shift in exploration philosophy, emphasizing sustainability and self-reliance.

Exploration is not without its challenges. The surface presents extreme conditions, from temperature swings of hundreds of degrees to radiation exposure in the absence of an atmosphere. Addressing these challenges requires innovative solutions, such as radiation shielding using regolith and energy storage systems capable of withstanding prolonged darkness during the two-week-long nights.

As exploration continues, it becomes clear that this celestial body is more than a scientific curiosity; it is a proving ground for humanity's aspirations. Each mission tests technologies and strategies that will be essential for venturing further into the solar system. From Mars to the asteroid belt, the lessons learned here will guide future generations of explorers.

The satellite's proximity to Earth has made it the first logical step in humanity's journey beyond its home planet, but its significance extends far beyond convenience. It is a partner in understanding the universe and a reminder of what can be achieved through ambition, collaboration, and perseverance. The discoveries made here will not only deepen our knowledge of the solar system but also inspire humanity to continue reaching for the stars.

Exploration has always been driven by a desire to understand and to expand. This is no less true for the quest to explore our closest celestial companion. While early robotic missions and human landings revolutionized scientific understanding, the motivations for continued exploration have evolved to include technological innovation, resource utilization, and preparation for humanity's future in space.

The surface, stark yet mesmerizing, offers both challenges and opportunities for exploration. Its regolith, a fine dust created by billions of years of impacts, provides a window into the solar system's history. However, it also presents hazards for equipment and astronauts. The abrasive nature of regolith has posed significant challenges, such as damaging seals, clogging joints, and wearing down materials. Addressing these issues has driven advancements in engineering, from dust-resistant technologies to robust filtration systems, ensuring future missions can withstand prolonged exposure.

One of the most tantalizing discoveries of recent years is the presence of volatiles, including water ice, in permanently shadowed regions near the poles. These areas, shielded from sunlight for billions of years, are among the coldest locations in the solar system. Here, water molecules delivered by comets, asteroids, and solar wind have accumulated, forming deposits that could be mined and processed for various uses. The ability to extract and utilize these resources is a cornerstone of modern exploration strategies.

The polar regions, once considered inhospitable, have become focal points for exploration planning. Their unique environmental conditions present both opportunities and obstacles. On the one hand, they harbor critical resources and offer the potential for near-continuous sunlight during their long summer days, ideal for solar power generation. On the other hand, navigating and operating in these regions requires specialized technologies capable of withstanding extreme temperatures and ensuring energy availability during prolonged periods of darkness.

Modern exploration programs increasingly rely on robotic precursors to scout potential landing sites, map resources, and test technologies. These missions, often conducted in partnership with international collaborators, are vital for reducing risks associated with human exploration. For example, rovers like NASA's VIPER are designed to prospect for water ice and assess its accessibility, providing critical data for planning sustainable human missions.

Sustainability has become a guiding principle of exploration. The concept of in-situ resource utilization, or ISRU, is central to this vision. By using local materials to produce essentials such as water, oxygen, and construction materials, future missions can reduce dependence on costly Earth-based resupply. This approach is not only practical but also necessary for long-term exploration goals, such as establishing permanent outposts and supporting interplanetary missions.

Infrastructure development is another key focus. Plans for lunar bases include habitats that can withstand radiation and micrometeorite impacts, systems for generating and storing energy, and life support systems capable of recycling air and water. These bases would serve as hubs for scientific research and staging points for missions to deeper destinations. Technologies tested here could later be adapted for harsher environments, such as those on Mars.

The prospect of building infrastructure raises broader questions about the future of exploration. How will nations and private entities share responsibilities and resources? What frameworks will govern the use of local materials? These questions reflect the need for international agreements that balance exploration with preservation and ensure equitable access to opportunities and benefits.

Beyond practical considerations, exploration continues to uncover scientific mysteries. Recent missions have revealed the presence of unusual geological features, such as pits and lava tubes, which could provide natural shelters for future habitats. These features, formed during ancient volcanic activity, offer a glimpse into the satellite's dynamic past and present intriguing possibilities for future use.

The journey of exploration is as much about discovery as it is about inspiration. Every mission, whether robotic or crewed, captures the imagination of millions, reminding humanity of its shared desire to understand and achieve. The images of Earth rising above the horizon, captured by Apollo astronauts, remain some of the most iconic and moving depictions of our planet. They highlight not only the uniqueness of our home but also the vastness of the universe and the possibilities that lie ahead.

As humanity stands on the cusp of a new era of exploration, the focus shifts from fleeting visits to establishing a lasting presence. This vision, while ambitious, is a natural extension of human ingenuity and resilience. By leveraging the lessons of the past and the technologies of the present, humanity moves closer to realizing its place as a multi-planetary species.

The exploration of our closest neighbor is far from complete. Every step taken—whether by robotic rover or human astronaut—is another chapter in a story that began with ancient stargazers and continues to unfold. The discoveries made today will shape the missions of tomorrow, pushing the boundaries of what is possible and inspiring generations to come.

The journey of exploring Earth's closest neighbor has progressed from fleeting robotic encounters to detailed studies aimed at understanding its resources, geology, and potential for human habitation. Each mission represents a building block in humanity's long-term vision of space exploration, driven by curiosity and the need to expand our horizons.

A critical aspect of exploration today is resource mapping and utilization. The discovery of water ice in polar regions has reshaped strategies, highlighting the importance of this resource for sustaining life and fueling spacecraft. Extracting water not only provides hydration and breathable oxygen but also yields hydrogen for rocket fuel. This ability to produce resources in situ has transformed perceptions of exploration, making long-term habitation feasible rather than merely speculative.

The exploration of permanently shadowed craters at the poles poses unique challenges. These regions are shrouded in darkness, with temperatures plummeting to extremes that test the limits of current technology. Robotic missions equipped with advanced sensors and autonomous navigation systems are essential for characterizing these areas. Future plans include deploying specialized equipment capable of drilling into icy deposits and processing materials in harsh conditions, paving the way for human operations.

Building sustainable infrastructure on the surface is another key focus. Concepts for habitats have evolved from rigid modules to flexible designs that incorporate local materials. Regolith, abundant on the surface, is being explored as a potential building material for shielding habitats from radiation and micrometeorite impacts. Additive manufacturing, or 3D printing, is a promising technique for creating structures directly on-site, reducing the need for heavy materials to be launched from Earth.

Energy generation and storage remain central to the success of long-term missions. The extended nights present challenges for solar power systems, prompting innovations in energy storage and backup solutions. Options such as fuel cells and small nuclear reactors are being developed to ensure a continuous power supply during periods of darkness. In sunlit regions near the poles, where sunlight can last for extended durations, solar arrays offer a reliable energy source, but these must be paired with systems capable of storing energy for when sunlight is unavailable.

Scientific exploration continues to uncover surprises about this celestial body. Recent discoveries include pits and caverns that may lead to underground lava tubes. These formations, remnants of ancient volcanic activity, could serve as natural shelters for habitats, offering protection from radiation and extreme temperatures. Their exploration may also reveal insights into its geological history, including clues about its thermal evolution and the forces that shaped its interior.

While technological advancements drive much of modern exploration, the human element remains at the heart of the endeavor. Astronauts bring adaptability, creativity, and the ability to make real-time decisions that robotic systems cannot yet replicate. Future missions envision astronauts not only conducting research but also building infrastructure, overseeing robotic operations, and testing systems for deeper space exploration.

As the vision of a permanent presence takes shape, international collaboration becomes increasingly vital. Programs like Artemis exemplify this spirit of partnership, bringing together nations and private companies to achieve shared goals. Such collaboration ensures that exploration benefits humanity as a whole while fostering innovation and cultural exchange. However, this cooperation also requires careful governance, with agreements that address the use of resources, environmental preservation, and equitable access.

The broader implications of exploration extend beyond practical achievements. It serves as a unifying endeavor that inspires innovation and fosters a sense of shared purpose. The iconic image of Earth rising above the horizon, taken during the Apollo missions, underscores the interconnectedness of life on our planet and the vast potential for discovery beyond it. Such moments remind humanity of its collective capacity to overcome challenges and reach new heights.

Looking ahead, this phase of exploration is not an endpoint but a foundation for future ambitions. The experience and knowledge gained will inform efforts to venture further into the solar system, with Mars as the next logical destination. The same principles of resource utilization, sustainable infrastructure, and international cooperation will guide these missions, building on the lessons learned.

In many ways, exploring this celestial body is a microcosm of humanity's journey into space. Each step forward represents a blending of curiosity, ingenuity, and resilience. The challenges encountered and overcome serve as a testament to what can be achieved when ambition meets determination. As we continue to explore, we not only unlock the secrets of another world but also refine our understanding of ourselves and our place in the cosmos.

The exploration of our nearest celestial neighbor is not merely a scientific or technological pursuit—it is a testament to humanity's enduring desire to venture into the unknown. Each mission, from the earliest robotic probes to the ambitious plans of the present, contributes to a growing body of knowledge and an expanding vision for the future. The challenges and triumphs encountered along the way highlight the complexity of turning what was once a distant dream into a tangible reality.

One of the most profound outcomes of exploration has been the revelation of just how interconnected Earth and its companion are. The samples returned during the Apollo missions unveiled a story of shared origins, with mineral compositions and isotopic similarities pointing to a violent past that forged the two bodies into a unique partnership. This understanding has redefined our place in the solar system, emphasizing the role that large impacts play in shaping planetary systems and their potential for hosting life.

These discoveries have inspired scientists and engineers to push boundaries, developing technologies to overcome the many obstacles presented by this harsh environment. Its surface, barren yet rich in resources, offers both opportunities and challenges. From the fine, abrasive regolith that clings to equipment to the extreme temperature variations that test materials, the surface demands innovative solutions for every aspect of exploration.

The prospect of utilizing local resources has shifted the paradigm of exploration. Where once every ounce of material had to be carried from Earth, the focus now is on sustainability and independence. The extraction of water ice for life support and fuel production is a cornerstone of this approach. Water, the most critical resource for human survival, can also be split into oxygen and hydrogen, providing breathable air and propellant for rockets. This capability transforms exploration from a transient endeavor to one that can support long-term habitation and travel to more distant destinations.

Building infrastructure capable of supporting human life is another critical challenge. Habitats must provide protection from radiation and micrometeorites while maintaining a stable internal environment. Advances in robotics and 3D printing are making it possible to construct shelters using regolith as a primary material. This approach minimizes the need for materials launched from Earth and represents a sustainable model for future exploration.

Energy systems are equally vital to success. The long nights, lasting roughly two weeks, present significant challenges for solar power systems. As a result, exploration efforts are incorporating a mix of energy technologies, including batteries, fuel cells, and compact nuclear reactors. These systems ensure a steady power supply for critical operations, from maintaining habitats to supporting scientific research.

Science remains at the heart of exploration, with each mission revealing new insights about this celestial body's past and present. Recent observations have identified volcanic features that suggest its geological activity may have persisted longer than previously thought. These findings raise questions about the thermal evolution of its interior and its potential for retaining heat-driven processes. Understanding these dynamics provides a window into the history of other rocky bodies in the solar system.

As exploration becomes increasingly international, collaboration has emerged as a defining feature of modern efforts. Programs like Artemis, the European Space Agency's contributions, and China's Chang'e missions demonstrate the power of pooling resources and expertise. This global approach not only accelerates progress but also fosters a sense of shared achievement, ensuring that exploration benefits all of humanity.

The exploration of this celestial neighbor also serves as a testing ground for humanity's ambitions beyond the solar system. The lessons learned in navigating its harsh environment, building sustainable systems, and managing complex international partnerships will guide future missions to Mars and beyond. By developing technologies and strategies here, humanity lays the groundwork for venturing further into the cosmos.

Perhaps most importantly, exploration continues to inspire. Each mission serves as a reminder of humanity's capacity to overcome challenges and achieve the extraordinary. The sight of astronauts walking on its surface, conducting experiments, and gazing back at Earth connects people across the globe, instilling a sense of unity and shared purpose. These moments are not just milestones in space exploration—they are milestones in human history.

As we stand at the threshold of a new era, the exploration of this celestial body is a reminder that the unknown is not something to be feared but embraced. It represents the possibilities that arise when curiosity, innovation, and collaboration converge. With each step forward, humanity not only deepens its understanding of the universe but also strengthens its resolve to explore the vast unknown.

Chapter 6: The Moon in Modern Science

Modern science has redefined humanity's understanding of the Moon, transforming it from a distant object of fascination into a vital subject of study with profound implications for planetary science, cosmology, and the future of exploration. Advances in technology and research have enabled scientists to explore its surface and interior in unprecedented detail, uncovering insights that inform not only our understanding of the Moon but also the dynamics of the solar system and beyond.

Central to this modern exploration is the growing realization of the Moon's role as a recorder of solar system history. Its surface, largely unaltered by weathering or tectonic activity, preserves a record of impacts and other events dating back billions of years. By studying the distribution and composition of craters, scientists can reconstruct the timeline of asteroid and comet activity during the chaotic early solar system. This information provides critical context for understanding planetary formation and the conditions that shaped Earth's own development.

Recent missions have expanded our knowledge of the Moon's composition, revealing a wealth of detail about its crust, mantle, and core. Data from orbiters and landers have identified variations in surface minerals, shedding light on the processes that shaped its geology. For example, the discovery of regions enriched in thorium and potassium has provided evidence for prolonged volcanic activity, challenging earlier assumptions about its thermal evolution. These findings suggest that the Moon remained geologically active for longer than previously thought, raising questions about how its interior cooled and whether similar processes might occur on other rocky worlds.

One of the most transformative discoveries in modern lunar science has been the detection of water in various forms. Initially considered a dry, barren world, the Moon is now known to harbor water ice in permanently shadowed craters near its poles. These deposits, believed to have accumulated over billions of years, are thought to be remnants of comet impacts, asteroid collisions, and interactions with the solar wind. This discovery has far-reaching implications, not only for understanding the Moon's history but also for its potential as a resource for future exploration.

The ability to utilize water ice as a resource has become a cornerstone of modern scientific and exploratory efforts. Water can be split into hydrogen and oxygen, providing both fuel and breathable air for astronauts. Its presence opens the door to long-term human habitation and the establishment of permanent outposts. Research into the distribution, purity, and accessibility of these deposits continues to shape plans for future missions, guiding the selection of landing sites and the development of extraction technologies.

The Moon's unique environment has also made it an ideal laboratory for studying fundamental scientific questions. Its lack of atmosphere and magnetic field exposes the surface to cosmic radiation and the solar wind, providing a pristine environment for studying these phenomena. Instruments placed on the surface have captured data on the behavior of high-energy particles, offering insights into the Sun's activity and the effects of radiation on planetary surfaces.

In addition to its value for planetary science, the Moon serves as a platform for observing the broader universe. Telescopes positioned on its surface would benefit from the absence of an atmosphere, enabling clearer and more detailed observations of distant stars and galaxies. The far side, shielded from Earth's radio emissions, is particularly well-suited for radio astronomy, offering an unparalleled opportunity to study the early universe and detect faint signals from distant celestial objects.

Modern science has also focused on understanding the Moon's potential for human habitation. Research into its regolith has revealed not only its challenges—such as its abrasive nature and tendency to cling to surfaces—but also its potential as a building material. Advances in 3D printing have demonstrated the feasibility of using regolith to construct habitats, shielding astronauts from radiation and extreme temperatures. These studies are critical for developing sustainable approaches to living and working on the Moon, ensuring that future missions can operate with minimal reliance on Earth-based resources.

The Moon's continued exploration has also fueled innovation in robotics and automation. Modern missions rely on highly sophisticated robotic systems to conduct research, map resources, and test technologies for future human operations. These robots are designed to navigate rugged terrain, drill into icy deposits, and operate autonomously for extended periods. The lessons learned from these missions are directly applicable to exploring more distant destinations, such as Mars and icy moons in the outer solar system.

As scientific research progresses, it is becoming increasingly clear that the Moon holds answers to questions that extend far beyond its surface. Its role in stabilizing Earth's axial tilt and influencing its climate provides insights into the conditions necessary for life on other planets. By studying the Earth-Moon system, scientists can refine models of planetary habitability, guiding the search for exoplanets with similar characteristics.

The Moon is not merely a relic of the past; it is a dynamic, evolving world that continues to shape our understanding of the cosmos. Its exploration is a testament to the power of science and the unyielding human spirit of discovery. As we continue to study its surface, interior, and environment, we uncover new layers of complexity, challenging our assumptions and expanding the boundaries of knowledge.

The Moon's role in modern science has expanded far beyond the confines of planetary geology. As research delves deeper into its properties, the Moon has become a focal point for understanding processes that impact planets, stars, and the broader cosmos.

Its environment, free from atmospheric interference and tectonic activity, offers unparalleled opportunities for conducting experiments and observing phenomena that cannot be easily replicated on Earth.

One area of focus has been the study of impact craters, which serve as a historical archive of the solar system's most violent events. The lack of erosion and tectonic activity means that craters, some billions of years old, remain well-preserved. By analyzing these features, scientists can infer the frequency and intensity of asteroid bombardments over time. This knowledge is critical for understanding how these impacts influenced planetary evolution, including the delivery of water and organic compounds that may have played a role in the emergence of life.

The distribution and composition of impact basins also provide insights into the Moon's interior. Craters reveal subsurface material, offering clues about the layering and composition of the crust and mantle. The South Pole–Aitken Basin, one of the largest and oldest impact features in the solar system, is of particular interest. Its immense size and depth expose material from deep within the crust, making it a prime target for future missions aimed at studying the Moon's geological history.

Advances in remote sensing technology have revolutionized the study of the Moon's surface. Instruments aboard orbiters, such as spectrometers and radars, have mapped the distribution of minerals and detected variations in surface composition. These data have revealed the presence of materials like ilmenite, a titanium-rich mineral that could be used to produce oxygen and extract valuable metals. Understanding the availability and distribution of such resources is essential for planning sustainable exploration and utilization efforts.

The discovery of water ice near the poles has transformed scientific priorities. Research into the origins and distribution of this ice has yielded new theories about the Moon's interaction with the solar wind and its ability to trap volatiles. These studies have broader implications, shedding light on how water and other key compounds are distributed across the solar system. They also raise intriguing possibilities about the role of small bodies, like comets and asteroids, in delivering these materials to planets.

Lunar regolith, once viewed solely as a challenge for exploration, is now considered a resource with vast potential. This fine, dusty material contains oxygen bound in minerals and trace amounts of hydrogen and helium. By developing technologies to process regolith, scientists hope to create systems for producing oxygen, water, and building materials directly on the Moon. This approach aligns with the broader goal of reducing dependence on Earth-based resources, making long-term habitation more feasible.

Another critical area of research is the Moon's magnetic anomalies. While it lacks a global magnetic field, localized regions exhibit significant magnetism, puzzling scientists for decades. These anomalies, thought to be remnants of an ancient dynamo or the result of large impacts, provide clues about the Moon's thermal and magnetic history. Understanding these features not only informs theories about the Moon's evolution but also offers insights into how magnetic fields develop and dissipate on other planetary bodies.

The Moon's environment has also proven ideal for conducting fundamental physics experiments. The absence of an atmosphere and the low-gravity conditions make it an excellent location for testing theories of gravity, relativity, and particle physics. Instruments deployed on the surface could measure cosmic phenomena with precision, free from the distortions caused by Earth's atmosphere. These experiments have the potential to unlock new dimensions of understanding, from dark matter and energy to the fundamental constants of nature.

In the realm of astronomy, the Moon's far side offers an unmatched opportunity for radio observations. Shielded from Earth's radio interference, this region is perfect for studying the early universe. Observatories placed here could detect faint signals from the first stars and galaxies, providing insights into the cosmic dawn. Additionally, the Moon's stability and lack of seismic activity make it an ideal location for building large-scale telescopes capable of capturing detailed images of distant celestial objects.

As exploration progresses, the integration of robotic and human missions is becoming a cornerstone of scientific advancement. Robots equipped with advanced sensors and AI are capable of conducting preliminary studies, identifying resources, and preparing sites for human arrival. These missions complement human capabilities, allowing astronauts to focus on complex tasks such as deploying scientific instruments, repairing equipment, and conducting experiments.

The Moon's proximity to Earth has also made it a valuable testbed for technologies intended for use on Mars and beyond. Systems for extracting resources, building habitats, and maintaining life support can be tested and refined here, where the risks are lower and resupply missions are feasible. This iterative process ensures that technologies are reliable and effective when deployed in more challenging environments.

Perhaps one of the most exciting aspects of modern lunar science is its potential to address questions about planetary habitability and the origins of life. By studying the Earth-Moon system, scientists can identify key factors that make a planet suitable for life, such as stable climates, the presence of water, and protection from harmful radiation. These lessons guide the search for exoplanets with similar characteristics, expanding the scope of astrobiology and the quest to understand life's place in the universe.

In many ways, the Moon serves as both a mirror and a gateway. Its surface reflects the processes that shaped the early solar system, offering a glimpse into Earth's own tumultuous history. At the same time, its exploration lays the groundwork for humanity's future in space, providing a stepping stone to the stars. Modern science continues to uncover its secrets, proving that even our closest neighbor still holds mysteries waiting to be solved.

Modern exploration has revealed that the Moon is not just a relic of the solar system's past but a dynamic environment with far-reaching implications for science and humanity's future. It serves as a unique laboratory for examining processes that shape planets, studying cosmic phenomena, and testing technologies crucial for interplanetary travel. As the focus of lunar science broadens, researchers are uncovering new dimensions of its significance.

One of the most critical areas of study is the Moon's role in understanding the early solar system. Its surface, untouched by erosion or plate tectonics, preserves a record of impacts that shaped the inner planets. By dating and analyzing these craters, scientists can reconstruct a timeline of asteroid activity, providing insights into the chaotic era of planetary formation. This research is particularly relevant for understanding the Late Heavy Bombardment, a period marked by intense collisions that likely influenced the development of Earth's crust and atmosphere.

The Moon's composition continues to reveal details about its origin and evolution. The presence of anorthosite in the highlands indicates that its early surface was dominated by a magma ocean, which cooled and solidified over millions of years. Meanwhile, basaltic plains in the maria suggest prolonged volcanic activity fueled by radioactive decay in the mantle. These findings challenge earlier views of the Moon as a geologically simple body, instead highlighting its dynamic history.

Recent studies have focused on the Moon's thermal evolution, particularly its partially molten core and mantle. Seismometers deployed during the Apollo missions detected moonquakes, providing clues about its internal structure. These data suggest that heat generated by the decay of radioactive elements played a significant role in driving early volcanic activity. Understanding these processes has broader implications, offering a comparative model for studying other rocky bodies in the solar system.

Water, once thought to be entirely absent from the Moon, has emerged as a central focus of modern research. Advances in spectroscopy have revealed not only water ice in polar regions but also traces of hydroxyl molecules bound to minerals across the surface. This discovery raises questions about the origins of lunar water, with theories ranging from delivery by comets and asteroids to chemical interactions with the solar wind. The distribution and accessibility of water are now key factors in planning future missions, as they hold the potential to support human habitation and resource utilization.

Beyond water, the Moon's surface materials contain valuable resources that could be harnessed for future exploration. Regolith, rich in oxygen and metals, is being studied for its potential to support construction and energy production. Techniques such as extracting oxygen from ilmenite or using microwaves to sinter regolith into building blocks are being tested to create sustainable infrastructure. These advancements align with the broader goal of developing self-sufficient systems for long-term human presence.

The Moon's environment also provides unique opportunities for studying cosmic radiation and solar activity. Without an atmosphere or magnetic field to shield it, the surface is directly exposed to high-energy particles from the Sun and beyond. Instruments deployed on robotic and human missions have captured data on radiation levels, which are critical for assessing the risks to astronauts and designing protective systems. Additionally, the Moon's surface acts as a natural detector, preserving a record of solar wind interactions that offer insights into the Sun's behavior over billions of years.

The far side of the Moon, shielded from Earth's radio emissions, is a prime location for advancing radio astronomy. This region provides a pristine environment for observing low-frequency radio waves, which are blocked by Earth's atmosphere. Future observatories placed here could study the early universe, detecting signals from the first stars and galaxies. Such research would complement other astronomical efforts, offering a new perspective on the cosmos.

The integration of robotics and human exploration is driving a new era of lunar science. Robotic missions are being used to map resources, test technologies, and prepare sites for human operations. These missions complement human capabilities, enabling astronauts to focus on tasks that require decision-making and adaptability. For example, robots can conduct preliminary drilling operations, while humans analyze samples and deploy more sophisticated instruments.

Lunar science is also contributing to the search for life beyond Earth. By studying the conditions that make Earth habitable, such as its stable climate and the presence of water, scientists can identify similar factors on exoplanets. The Moon's role in stabilizing Earth's axial tilt and influencing its tides underscores the importance of moons in shaping planetary environments. These lessons guide the search for habitable worlds, highlighting the interconnectedness of planetary systems.

As exploration progresses, the Moon remains a proving ground for technologies and strategies aimed at Mars and beyond. Systems for resource extraction, habitat construction, and life support are being tested in its challenging environment, ensuring their reliability in more distant and inhospitable locations. This iterative process allows researchers to refine their designs, building confidence for future interplanetary missions.

The Moon's significance extends beyond science and exploration—it is a source of inspiration that unites humanity in the pursuit of knowledge. The milestones achieved here, from the first steps on its surface to the ongoing discoveries of robotic missions, remind us of our capacity to innovate and collaborate. As we continue to study the Moon, we deepen our understanding of the universe and reaffirm our commitment to pushing the boundaries of what is possible.

Modern science has turned the Moon into one of the most thoroughly studied celestial objects, yet it continues to surprise researchers with its complexity and relevance. As missions probe deeper into its mysteries, the Moon reveals insights not only about its own history but also about the processes that shape planets and systems throughout the galaxy.

One of the most intriguing aspects of modern lunar research is its role as a natural laboratory for studying planetary impacts. Craters across its surface preserve the record of billions of years of collisions, from small meteorites to massive asteroids. These impacts provide clues about the intensity and frequency of such events during the early solar system. By understanding these patterns, scientists can extrapolate the risks of future impacts on Earth and other planets. This research has direct implications for planetary defense, offering strategies to mitigate potential threats from near-Earth objects.

The Moon's interior has become a key focus of study, particularly its partially molten core and the remnants of its magnetic field. Seismic data, collected during the Apollo era, indicate that its interior is layered, with a solid inner core surrounded by a fluid outer layer. This configuration, though less dynamic than Earth's core, offers a window into how planetary bodies cool and lose their magnetic fields. Understanding these processes is crucial for interpreting the histories of other celestial bodies, such as Mars, which also exhibits signs of a once-active magnetic field.

Recent studies of lunar surface chemistry have revealed the presence of trace elements and volatiles that were previously unexpected. For example, the detection of hydroxyl molecules across the surface suggests a more active chemical environment than initially believed. These molecules may be formed through interactions between the solar wind and surface minerals, highlighting the dynamic interplay between the Moon's surface and its surrounding space environment.

The discovery of water ice in polar craters has been a game-changer for both science and exploration. These deposits, likely accumulated from comet impacts and solar wind contributions, offer a snapshot of the materials that have shaped the inner solar system. They also represent a critical resource for future missions, as water can be used for drinking, oxygen production, and rocket fuel. Developing the technology to extract and process this ice is a central goal of many current and upcoming missions, bridging the gap between scientific discovery and practical application.

The Moon's surface is also a pristine environment for studying solar and cosmic radiation. Without the protective shield of an atmosphere or magnetic field, the surface is directly exposed to high-energy particles. Instruments deployed by recent missions have measured these radiation levels, providing vital data for understanding the risks to astronauts and designing effective shielding for habitats. This research extends beyond the Moon, informing strategies for human exploration of Mars and other distant destinations.

In the field of astronomy, the Moon's far side has emerged as a potential site for transformative discoveries. Shielded from Earth's radio signals, this region offers an unparalleled opportunity for low-frequency radio observations. Future telescopes placed here could detect faint signals from the earliest periods of cosmic history, revealing how the first stars and galaxies formed. These efforts complement ongoing astronomical research, providing a unique perspective on the universe's evolution.

The use of robotics has been pivotal in advancing lunar science. Modern rovers and landers are equipped with sophisticated instruments capable of conducting experiments, analyzing materials, and mapping the surface in high resolution. For example, China's Chang'e-4 mission, which landed on the far side, has provided unprecedented data on its geology and environment. These robotic missions lay the groundwork for human exploration, identifying resources, and testing technologies that will be critical for future missions.

Human missions, while fewer in number, have contributed immeasurably to our understanding of the Moon. The Apollo program, with its detailed surface studies and sample returns, remains the gold standard of lunar science. The samples collected during these missions continue to yield new discoveries, thanks to advancements in analytical techniques. For instance, modern instruments have detected isotopic variations in lunar rocks that provide clues about its formation and the nature of the impact that created it.

As humanity prepares for its return to the Moon, the focus has shifted from short-term visits to establishing a lasting presence. This vision includes constructing habitats, harvesting resources, and conducting scientific research on a scale never before attempted. The lessons learned from these efforts will extend beyond the Moon, shaping the strategies and technologies needed for exploring more distant worlds.

The Moon also plays a role in understanding Earth itself. Its influence on tides, axial stability, and climate provides a framework for studying similar dynamics on other planets. By examining this relationship, scientists can refine models of planetary evolution and identify the conditions that make a planet habitable. These insights are critical for the search for life beyond the solar system, guiding efforts to identify exoplanets with Earth-like characteristics.

In many ways, the Moon represents the intersection of science and exploration. Its study deepens our understanding of the universe, while its exploration drives technological innovation and international collaboration. Each discovery brings us closer to answering fundamental questions about our origins and our future, reaffirming the Moon's place as a cornerstone of modern science.

The Moon's role in modern science transcends its immediate relationship with Earth, emerging as a key player in broader questions about the nature of planetary systems and the universe itself. It acts as a bridge between past and future, offering a preserved record of solar system history while serving as a foundation for humanity's aspirations in space exploration.

A key area of study continues to be the Moon's surface and its unique environment. The regolith, though deceptively simple in appearance, is a treasure trove of information. Every layer of this fine, dusty material tells a story about cosmic impacts, solar activity, and interactions with interplanetary particles. Recent advancements in spectroscopy and remote sensing have allowed researchers to identify subtle variations in its composition, providing insights into the processes that shaped the Moon over billions of years.

Impact craters, among the most prominent features on the surface, are windows into the Moon's geological and cosmological history. By studying the distribution and structure of craters, scientists can trace the history of asteroid and comet activity across the inner solar system. These studies not only illuminate the Moon's past but also offer critical data for understanding the risks posed by near-Earth objects and developing strategies for planetary defense.

The Moon's polar regions have become a focal point for exploration, both robotic and human. Permanently shadowed craters, some of the coldest places in the solar system, harbor water ice and other volatiles that have accumulated over billions of years. These deposits are invaluable for understanding the sources and delivery mechanisms of water in the solar system. Moreover, their potential for supporting life support systems and rocket fuel production makes them essential to the future of sustainable exploration.

The development of in-situ resource utilization (ISRU) technologies has transformed how scientists and engineers approach lunar exploration. Techniques for extracting oxygen, water, and metals from regolith and ice deposits are advancing rapidly, bringing the concept of self-sufficient lunar outposts closer to reality. Such capabilities will not only enable long-term habitation but also reduce the logistical challenges and costs of supporting missions to more distant destinations like Mars.

Energy production and storage remain critical challenges for long-term operations. The two-week-long lunar nights require systems capable of storing energy efficiently and providing continuous power. Solar arrays paired with advanced battery systems are a promising solution, while small modular reactors are being considered as a reliable alternative. These innovations are not only vital for the Moon but also represent technologies that could be adapted for use in other environments.

In the realm of planetary science, the Moon serves as a comparative model for understanding other rocky bodies. Its lack of atmosphere and plate tectonics allows scientists to study surface features and geological processes that have been erased on Earth. For example, the study of ancient volcanic activity on the Moon has provided insights into the cooling and differentiation of planetary interiors, helping to refine models of planetary evolution across the solar system.

Astronomical research is another domain where the Moon's unique environment offers distinct advantages. The far side, free from Earth's radio interference, is an ideal location for radio astronomy. Observatories placed here could detect low-frequency signals from the early universe, offering a glimpse into the formation of the first stars and galaxies. This region could also serve as a platform for studying exoplanets, searching for biosignatures, and observing distant phenomena with unparalleled clarity.

The Moon's proximity to Earth has made it a valuable testbed for technologies and strategies aimed at interplanetary travel. Systems for habitat construction, life support, and resource utilization are being tested and refined in its challenging environment. These efforts ensure that missions to more distant destinations are built on a foundation of proven techniques and technologies, minimizing risks and maximizing the chances of success.

Human exploration remains a cornerstone of lunar science. The return of humans to the surface, planned under initiatives like NASA's Artemis program, represents a significant leap forward. Astronauts will bring adaptability and problem-solving skills that complement robotic capabilities, enabling more complex and nuanced investigations. These missions aim to establish a sustainable presence, laying the groundwork for scientific discovery and broader exploration efforts.

The Moon also holds cultural and philosophical significance. Its exploration unites humanity in a shared pursuit of knowledge, transcending national and cultural boundaries. Each mission serves as a reminder of what can be achieved through collaboration and innovation. The iconic image of Earth rising above the lunar horizon, captured during the Apollo missions, continues to inspire a sense of global unity and purpose.

As science and exploration progress, the Moon remains at the forefront of humanity's efforts to understand the cosmos. It challenges us to think beyond our immediate needs and consider the long-term implications of our actions and discoveries. From unlocking the secrets of the solar system's formation to preparing for life on other planets, the Moon is a constant reminder of the possibilities that lie ahead.

Modern science views the Moon not just as an object of study but as a partner in humanity's journey to the stars. Its exploration is a testament to our capacity for ingenuity and our enduring curiosity about the universe. As we continue to unlock its mysteries, the Moon will undoubtedly play a central role in shaping the next chapter of human history.

Chapter 7: Living and Working on the Moon

As humanity takes its next steps in space exploration, the prospect of living and working on the Moon transitions from science fiction to a tangible goal. Establishing a sustainable presence on its surface presents unique challenges but also unprecedented opportunities for scientific discovery, technological innovation, and interplanetary exploration.

A permanent or semi-permanent lunar presence hinges on the ability to adapt to an environment unlike any encountered on Earth. The Moon's lack of atmosphere, extreme temperature fluctuations, and exposure to cosmic radiation demand innovative solutions to support human life. At the same time, its proximity to Earth and wealth of resources make it an ideal proving ground for the technologies and systems needed for long-term space exploration.

The Moon's environment is both harsh and unforgiving. Temperatures swing dramatically between day and night, ranging from over 100 degrees Celsius during the day to below -150 degrees Celsius at night. These extremes necessitate habitats capable of maintaining a stable internal climate while withstanding the thermal stress imposed by the external environment.

Radiation is another critical concern. Without an atmosphere or magnetic field to provide shielding, the surface is bombarded by solar and cosmic radiation. Prolonged exposure can increase the risk of cancer and other health issues, making radiation protection a top priority for habitat design. Potential solutions include covering habitats with regolith to provide natural shielding or using advanced materials that block radiation.

Dust, or regolith, poses its own set of challenges. This fine, abrasive material clings to surfaces, infiltrates machinery, and poses respiratory hazards if inhaled. Developing effective dust mitigation strategies, such as electrostatic barriers or coatings, is essential for maintaining the integrity of equipment and ensuring the safety of astronauts. The first step toward living on the Moon is constructing habitats capable of supporting human life. These structures must provide protection from radiation, micrometeorites, and temperature extremes while meeting the needs of astronauts for air, water, food, and waste management. Advances in 3D printing technology are paving the way for habitats constructed using local materials. By using regolith as a building material, these habitats can be built on-site, reducing the need to transport heavy components from Earth.

Energy generation and storage are critical for sustaining life on the Moon. Solar power is a promising solution, particularly in polar regions where sunlight is available for extended periods. However, energy storage systems are necessary to provide power during the two-week-long lunar nights. Batteries, fuel cells, and small modular reactors are all being explored as viable options.

Life support systems on the Moon must be highly efficient and self-sufficient. Recycling air and water is essential, given the logistical challenges of resupply from Earth. Technologies developed for the International Space Station (ISS) provide a foundation for these systems, but they will need to be adapted for the Moon's environment and scaled up for longer missions.

Food production is another critical consideration. While initial missions may rely on pre-packaged supplies, long-term habitation requires the ability to grow food on-site. Experiments on the ISS have demonstrated the feasibility of growing plants in microgravity, but the Moon's low gravity presents unique challenges. Greenhouses equipped with LED lighting and hydroponic systems are being designed to test food production in lunar conditions.

A key aspect of living and working on the Moon is utilizing its resources to reduce reliance on Earth. Water ice, discovered in polar craters, is one of the most valuable resources. Extracting and processing this ice can provide drinking water, breathable oxygen, and hydrogen for rocket fuel. Technologies for mining and refining these resources are a top priority for future missions.

Regolith, abundant across the surface, is another critical resource. In addition to serving as a construction material, regolith contains oxygen bound in minerals. Techniques for extracting oxygen from regolith are being developed, offering a way to produce breathable air on-site. These efforts align with the broader goal of creating self-sustaining systems that support long-term habitation.

Living on the Moon is not just about survival—it's also about productivity. Scientific research, resource extraction, and infrastructure development will be central to lunar operations. The low-gravity environment, while challenging, also offers unique opportunities for innovation. Heavy equipment can be moved more easily, and new manufacturing techniques can be developed to take advantage of the reduced gravity.

Robots will play a crucial role in supporting human activities on the Moon. Autonomous systems equipped with AI and advanced sensors can conduct preliminary surveys, transport materials, and perform maintenance tasks. These robots will complement human efforts, enabling astronauts to focus on more complex operations. Isolation, confinement, and the lack of natural environments pose significant psychological challenges for astronauts living on the Moon. Maintaining mental health is as important as ensuring physical well-being. Strategies for addressing these challenges include designing habitats with communal and private spaces, providing access to virtual reality environments that simulate Earth-like settings, and implementing regular communication with friends and family back home.

Establishing a presence on the Moon is not an end in itself but a stepping stone toward broader exploration goals. The technologies and systems developed here will serve as a foundation for missions to Mars and beyond. The Moon's proximity to Earth makes it an ideal testing ground, allowing scientists and engineers to refine their approaches before deploying them in more distant and challenging environments.

The prospect of living and working on the Moon is a bold vision, but it is one grounded in the progress of modern science and technology. Each step forward represents a leap toward understanding our place in the universe and unlocking the potential for humanity to thrive beyond Earth.

The vision of establishing a permanent or semi-permanent human presence on the Moon is no longer a distant dream. Advances in science, technology, and international collaboration are converging to address the challenges of lunar living. The unique environment offers a proving ground for the development of systems that will one day support humanity's ventures deeper into the solar system.

Building infrastructure on the Moon is a multifaceted endeavor that begins with the establishment of a secure, adaptable base. Initial habitats will likely be modular structures designed to be transported from Earth, offering reliable life support and protection against environmental hazards. However, as technologies for in-situ resource utilization (ISRU) advance, these imported systems will evolve into more robust, locally constructed settlements.

Construction on the Moon leverages the abundance of regolith, which can be processed into bricks or sintered into solid structures using heat or microwaves. 3D printing technology is central to this effort, enabling the rapid assembly of shelters and infrastructure while minimizing material waste. These techniques reduce the reliance on Earth-based resources and allow for scalability as lunar operations expand.

Energy infrastructure is equally critical. Solar power, particularly in the polar regions, is the most viable source of renewable energy. Near the poles, some areas receive almost continuous sunlight, ideal for solar arrays. However, energy storage systems must be designed to provide power during extended lunar nights. Battery technology, fuel cells, and even nuclear reactors are being explored to meet these needs.

Sustainability is the cornerstone of lunar habitation. Life support systems must be designed to recycle air and water with near-perfect efficiency. Advanced systems, like those used on the International Space Station (ISS), will serve as prototypes, but they will need to be scaled up and adapted for long-term use in the Moon's low-gravity environment.

Water recycling is critical, as resupply missions from Earth are expensive and logistically complex. Every drop of water used in drinking, hygiene, and food production must be captured, purified, and reused. Technologies for extracting water from ice deposits in polar craters complement these recycling systems, ensuring a steady supply of this vital resource.

Oxygen generation is another key component of life support. Methods for extracting oxygen from regolith, which contains oxides, are under development. Electrolysis of water ice is also being tested as a means of producing breathable air. These systems not only sustain life but also provide oxygen for fuel production, reducing the need to transport propellant from Earth.

Long-term habitation requires reliable food sources. While early missions will rely on pre-packaged meals, sustainable lunar living depends on the ability to grow food on-site. The Moon's low gravity presents challenges for agriculture, including how plants uptake water and nutrients. However, experiments conducted on the ISS have demonstrated that plants can adapt to non-Earth environments.

Greenhouses designed for the Moon will use hydroponic or aeroponic systems to grow crops without soil. These systems maximize resource efficiency, recycling water and nutrients to sustain plant growth. LED lighting tailored to the specific needs of crops provides optimal conditions for photosynthesis.

Researchers are also exploring the use of regolith as a growth medium after processing to remove harmful compounds.

In addition to providing sustenance, agricultural systems play a role in maintaining mental well-being. The presence of living plants can help mitigate the psychological effects of isolation and confinement, offering astronauts a connection to nature in an otherwise sterile environment.

The Moon's resources are key to reducing dependence on Earth and enabling sustainable operations. Water ice in polar craters is the most critical resource, serving as the basis for life support and fuel production. Extracting this ice requires specialized technologies capable of operating in extremely cold and dark conditions. Robotic systems equipped with drills and heaters are being developed to mine and process these deposits efficiently.

Regolith is another valuable resource. In addition to construction, it can be used to produce oxygen through chemical processes. The extraction of metals, such as aluminum and titanium, is also being explored, offering materials for manufacturing tools, equipment, and infrastructure.

The presence of helium-3, a rare isotope, has attracted interest as a potential fuel for nuclear fusion. While the technology for helium-3 fusion is still in development, its abundance on the Moon represents a long-term opportunity for energy production.

Living on the Moon is not just about survival but also about achieving specific objectives. Research, construction, and maintenance will occupy much of an astronaut's time. The low-gravity environment allows for unique opportunities, such as moving heavy equipment with relative ease and conducting experiments that would be impossible on Earth.

Robotic systems will complement human efforts, taking on repetitive or hazardous tasks. Autonomous rovers can transport materials, conduct surveys, and perform maintenance, freeing astronauts to focus on complex activities. These systems are equipped with AI and machine learning capabilities, enabling them to adapt to the dynamic conditions of the lunar surface.

The physical and mental health of astronauts is paramount in lunar missions. Prolonged exposure to low gravity can lead to muscle atrophy and bone density loss. Exercise regimens, similar to those used on the ISS, are essential for mitigating these effects. Equipment such as resistance machines and treadmills will be integral to maintaining physical fitness.

Psychological well-being is another critical consideration. Isolation, confinement, and the monotony of the environment can take a toll on mental health. Strategies to address these challenges include designing habitats with communal and private spaces, providing regular communication with loved ones, and using virtual reality to simulate Earth-like environments.

The experience of living and working on the Moon will have implications far beyond its surface. It serves as a testbed for the technologies and systems needed for interplanetary missions. The lessons learned here will inform efforts to establish bases on Mars and potentially other moons and asteroids.

Moreover, the pursuit of lunar habitation inspires innovation and collaboration on a global scale. The challenges of living on the Moon drive advancements in engineering, robotics, and medicine that have applications on Earth. International partnerships ensure that exploration benefits humanity as a whole, fostering a spirit of cooperation in addressing shared challenges.

The Moon represents both a destination and a gateway. Establishing a presence here is not just about exploring a neighboring world—it is about preparing for the next steps in humanity's journey into the cosmos. As we overcome the challenges of lunar living, we lay the foundation for a future where humanity can thrive beyond Earth.

Chapter 8: Humanity's Next Great Economy

The Moon, long considered a scientific and exploratory frontier, is increasingly viewed as a cornerstone of humanity's future economic expansion. As interest in space exploration grows, so does the recognition of its potential to fuel a new era of economic activity. From resource extraction to infrastructure development, the Moon offers opportunities that could transform global industries and establish a space-based economy.

The Moon's surface and subsurface hold a wealth of resources that are critical for supporting human activity in space. Water ice, found in polar regions, is among the most valuable. Beyond its use for sustaining life, water can be split into hydrogen and oxygen, providing propellants for rockets. This capability reduces the cost and complexity of space missions by enabling refueling in space rather than relying on Earth-based launches.

In addition to water, the Moon contains significant quantities of metals such as aluminum, titanium, and iron. These materials can be used to construct habitats, infrastructure, and spacecraft components, creating a self-sustaining supply chain for space development. Advances in mining and processing technologies are key to unlocking these resources, ensuring they can be extracted and utilized efficiently.

Another highly sought-after resource is helium-3, a rare isotope embedded in the lunar regolith. Helium-3 has potential as a fuel for nuclear fusion, a technology that promises to provide clean and virtually limitless energy. While fusion power is still in the experimental stage, the Moon's reserves of helium-3 could one day be a cornerstone of a global energy revolution.

Building infrastructure on the Moon is a critical step toward establishing an economic presence. This includes habitats, energy systems, and transportation networks that support human activity and facilitate resource extraction. Technologies such as 3D printing using regolith are enabling the construction of durable structures directly on the Moon, reducing the need for materials to be launched from Earth.

Energy generation and storage are central to infrastructure development. Solar arrays, particularly in polar regions with prolonged sunlight, provide a renewable source of power. Coupled with advanced battery systems and small modular reactors, these energy systems ensure the continuous operation of habitats, mining equipment, and communication systems.

Transportation networks, including lunar rovers and landers, are also vital for economic activity. These vehicles enable the movement of materials and personnel across the surface, supporting both exploration and industrial operations. Autonomous systems, equipped with AI, are being developed to manage these networks efficiently.

The Moon's low-gravity environment offers unique opportunities for manufacturing processes that are challenging or impossible on Earth. For example, the production of certain materials, such as high-quality crystals and specialized alloys, may benefit from the reduced gravity and vacuum conditions. These materials could be used in advanced electronics, medical devices, and aerospace applications.

In addition to materials production, the Moon is an ideal location for assembling and launching spacecraft. Its low gravity significantly reduces the energy required for launches, making it a cost-effective hub for interplanetary missions. Spaceports on the Moon could support missions to Mars, asteroids, and beyond, establishing it as a critical node in a growing space economy.

Lunar tourism, while currently a niche concept, has the potential to become a significant industry as technology advances and costs decrease. Companies are already planning commercial missions to orbit the Moon or land on its surface, offering experiences for private citizens. The Moon's unique environment, from its stark landscapes to the breathtaking view of Earth, makes it an unparalleled destination for adventure and exploration.

Other commercial ventures, such as the establishment of research facilities and entertainment projects, could also contribute to a lunar economy. Collaborative efforts between governments, private companies, and international organizations are key to ensuring that these ventures align with broader goals of sustainability and equity.

The development of a lunar economy requires a framework for international collaboration and governance. Agreements like the Artemis Accords emphasize the importance of peaceful exploration, resource sharing, and sustainability. As nations and private entities expand their presence on the Moon, these frameworks will need to address complex issues such as property rights, resource allocation, and environmental protection.

Collaboration between governments and private companies is also essential for funding and advancing lunar projects. Public-private partnerships leverage the strengths of both sectors, combining governmental oversight and private sector innovation. This model is already proving successful in Earth-based industries and has the potential to drive significant progress in lunar development. While the potential of a lunar economy is immense, it is not without challenges. The high costs of transportation, the complexity of resource extraction, and the need for advanced infrastructure pose significant barriers. Addressing these challenges requires continued investment in research and development, as well as collaboration across industries and nations.

The opportunities, however, far outweigh the obstacles. By establishing a sustainable economic presence on the Moon, humanity can unlock resources that support long-term exploration, drive technological innovation, and create new industries. These efforts will not only benefit those directly involved but also have far-reaching impacts on global economies, inspiring new approaches to energy, manufacturing, and collaboration. The development of a lunar economy is not an endpoint but a stepping stone. The Moon's proximity to Earth and its abundance of resources make it an ideal starting point for humanity's expansion into the solar system. By mastering the challenges of living and working on the Moon, we gain the knowledge and capabilities needed to explore Mars, asteroids, and beyond.

As humanity takes its first steps toward a space-based economy, the Moon stands as a symbol of possibility and progress. Its development represents a convergence of science, industry, and ambition, laying the foundation for a future where the resources of space are harnessed to benefit all of humanity. The development of a lunar economy is poised to reshape how humanity approaches both space and terrestrial challenges. While the prospect of resource extraction, manufacturing, and innovation on the Moon may seem futuristic, the groundwork is already being laid through ongoing missions, technological advancements, and strategic partnerships. This economy is not just about exploring space—it's about rethinking economic systems, sustainability, and humanity's relationship with resources.

Water ice, found in abundance in permanently shadowed regions, is the cornerstone of any viable lunar economy. Its extraction and utilization have implications that extend far beyond the Moon. Water serves multiple roles: as a direct resource for human survival, a medium for oxygen and hydrogen production, and a key component in creating rocket fuel. The ability to refuel spacecraft on the Moon could transform how missions are planned, enabling longer durations and greater distances.

The process of water extraction involves technologies such as heating or chemical reactions to release and collect water vapor. These methods must be efficient and scalable, capable of operating in extreme temperatures and darkness. The infrastructure to process and store water represents one of the first critical investments in developing a lunar economy.

Beyond water, the Moon's surface contains a wealth of volatiles, such as carbon dioxide and methane, which can be used to create fuel, plastics, and other materials. These compounds, while present in trace amounts, could complement water-derived resources, expanding the range of applications for lunar materials.

Mining operations on the Moon focus on extracting valuable metals and minerals, including aluminum, titanium, and iron, which are critical for construction and manufacturing. The Moon's regolith also contains rare earth elements, which are essential for modern technologies like smartphones, batteries, and renewable energy systems. While their concentration is lower than on Earth, the lack of environmental and societal restrictions on the Moon makes their extraction an attractive prospect.

Processing lunar materials requires advanced techniques that minimize energy consumption and waste. One promising approach is the use of solar furnaces, which concentrate sunlight to smelt and refine metals. This method aligns with the principles of sustainability, reducing reliance on Earth-based resources while minimizing the environmental footprint of mining operations.

As mining and resource processing progress, the next logical step is the construction of infrastructure that supports both human habitation and industrial activity. Habitats, storage facilities, and transportation hubs are essential for creating a functional lunar base. The Moon's regolith offers a readily available construction material, which can be processed into bricks or 3D-printed into more complex structures.

In addition to physical structures, communication networks are crucial for coordinating activities on the Moon. High-bandwidth systems linking the Moon to Earth ensure seamless data transfer, enabling real-time control of robotic systems and supporting scientific and commercial endeavors. These networks will also facilitate communication between lunar outposts, fostering collaboration and efficiency.

Energy is the backbone of any economic activity, and the Moon's unique environment presents both challenges and opportunities for energy generation. Solar power remains the most viable renewable source, particularly in regions with prolonged sunlight. Advances in photovoltaic technology are enabling more efficient solar panels that can operate under the harsh conditions of space.

Energy storage systems are equally important, providing power during the long lunar nights. Batteries with enhanced capacities, fuel cells, and compact nuclear reactors are all under development to meet these demands. The ability to generate and store energy locally reduces dependence on Earth, ensuring the resilience and scalability of lunar operations.

The Moon's low gravity opens up new possibilities for manufacturing processes that are difficult or impossible on Earth. For instance, the production of ultra-pure crystals and specialized alloys could revolutionize industries like electronics and aerospace. These materials, produced in the Moon's unique conditions, could be exported to Earth or used to build more advanced systems for space exploration.

Additive manufacturing, or 3D printing, is another area of significant potential. By using local materials, this technology enables the production of tools, components, and even entire structures on demand. This capability not only reduces costs but also ensures that missions can adapt to changing circumstances, creating a more flexible and sustainable system.

As infrastructure develops, tourism emerges as a viable industry on the Moon. Initial ventures may involve orbital flights offering breathtaking views of Earth and the lunar surface. Over time, surface expeditions and extended stays in lunar habitats could become possible, attracting adventurers and space enthusiasts willing to experience the unique environment.

Beyond tourism, other commercial opportunities include film production, scientific research facilities, and corporate-sponsored missions. The Moon's iconic landscapes and cultural significance make it an appealing destination for projects that combine science, art, and public engagement.

The creation of a lunar economy will not only transform space but also drive innovation on Earth. Technologies developed for lunar operations, such as advanced robotics, energy systems, and resource extraction methods, have direct applications in industries like construction, renewable energy, and manufacturing. The challenges of living and working on the Moon inspire solutions that benefit humanity.

Additionally, the development of lunar resources could reduce the environmental pressures on Earth. By sourcing rare materials and energy in space, humanity can lessen the impact of terrestrial mining and energy production, contributing to global sustainability goals. The rise of a lunar economy is inherently a global endeavor, involving governments, private companies, and international organizations. Collaborative efforts, such as the Artemis program and the European Space Agency's contributions, demonstrate the benefits of pooling resources and expertise. At the same time, competition among nations and corporations drives innovation, pushing the boundaries of what is possible.

To ensure that the lunar economy develops equitably and sustainably, international governance frameworks are essential. Agreements like the Artemis Accords provide a foundation for resource sharing, environmental protection, and peaceful exploration. These frameworks will need to evolve as activity on the Moon increases, balancing the interests of stakeholders while preserving the Moon's unique environment.

The development of a lunar economy is not just an end in itself—it is a stepping stone to humanity's broader expansion into the solar system. The Moon's resources and infrastructure provide a foundation for missions to Mars, asteroids, and beyond. By mastering the challenges of lunar operations, humanity gains the knowledge and experience needed to explore deeper into space.

This vision of a space-based economy is not without its challenges, but the rewards are immense. The Moon offers a unique opportunity to redefine how humanity approaches resources, technology, and sustainability, setting the stage for a future where the possibilities are as vast as the universe itself.

Chapter 9: The Future of Space Exploration

The Moon, once the ultimate symbol of human aspiration, now serves as the foundation for humanity's next great leap into the cosmos. The lessons learned from its exploration, resource utilization, and habitation are shaping the strategies for reaching further into the solar system and beyond. The future of space exploration is driven by a combination of scientific curiosity, technological innovation, and the innate human desire to explore the unknown.

The logical progression from the Moon is Mars, a planet that has captured the imagination of scientists, engineers, and the public alike. The Moon provides a critical testing ground for the technologies and systems that will enable humans to live and work on another planet. The challenges of lunar exploration—resource extraction, habitat construction, and life support—mirror those of Mars, albeit on a smaller scale.

Mars presents a more complex set of challenges, including its thin atmosphere, colder climate, and greater distance from Earth. However, its potential for supporting life and its similarity to early Earth make it a prime target for exploration. Technologies developed on the Moon, such as in-situ resource utilization (ISRU) and advanced habitat systems, will be directly applicable to Martian missions.

The Moon's low gravity makes it an ideal location for assembling and launching spacecraft destined for Mars. By refueling and staging missions from the Moon, humanity can reduce the costs and complexities of interplanetary travel. The establishment of lunar infrastructure is thus a critical step in realizing the dream of reaching the Red Planet.

While Mars is the immediate focus of interplanetary exploration, the solar system offers countless other destinations. Asteroids, with their wealth of resources and scientific value, are among the most promising. These bodies contain water, metals, and organic compounds that could support both exploration and industrial activity. Missions to study and mine asteroids are already being planned, with the Moon serving as a base for staging these efforts.

The outer planets and their moons represent the next frontier. Jupiter's moon Europa and Saturn's moon Enceladus, both believed to harbor subsurface oceans, are prime candidates in the search for extraterrestrial life. The icy surfaces of these moons protect their liquid interiors, where geothermal energy could support microbial ecosystems. Robotic missions will pave the way for eventual human exploration, advancing our understanding of these distant worlds.

The future of space exploration extends beyond our solar system. Advances in propulsion technology, such as nuclear thermal propulsion and solar sails, are opening new possibilities for interstellar travel. Concepts like Breakthrough Starshot, which aims to send tiny probes to the Alpha Centauri system, demonstrate the potential for reaching other star systems within a human lifetime.

While interstellar travel remains speculative, the exploration of exoplanets is well underway. Telescopes like the James Webb Space Telescope (JWST) are providing detailed observations of planets orbiting other stars, identifying candidates that may be capable of supporting life. The knowledge gained from exploring the Moon and Mars will inform humanity's search for habitable worlds beyond our solar system.

Technology is the driving force behind the future of space exploration. Advances in robotics, artificial intelligence, and materials science are enabling missions that were once impossible. Autonomous systems are becoming increasingly capable, performing complex tasks with minimal human intervention. These technologies will be critical for exploring environments that are too hostile or distant for human presence.

Propulsion technology is also evolving, with innovations that promise faster and more efficient travel. Nuclear propulsion, in particular, has the potential to significantly reduce travel times to Mars and beyond. Other concepts, such as electric propulsion and antimatter engines, are being explored for their long-term potential.

Communication systems are another area of focus. As missions venture further from Earth, maintaining reliable communication becomes increasingly challenging. Advances in laser communication and quantum encryption are addressing these challenges, ensuring secure and efficient data transfer across vast distances.

The future of space exploration is a collaborative effort, involving governments, private companies, and international organizations. Programs like Artemis demonstrate the power of partnerships in achieving ambitious goals. By sharing resources and expertise, nations can overcome the immense costs and complexities of interplanetary exploration.

The private sector plays a growing role in space exploration, bringing innovation and efficiency to the field. Companies like SpaceX, Blue Origin, and Rocket Lab are developing new technologies and reducing the cost of access to space. These efforts complement governmental initiatives, accelerating progress and opening new possibilities for exploration and commercialization.

As humanity expands its presence in space, ethical considerations become increasingly important. Questions about resource utilization, planetary protection, and the preservation of space environments must be addressed. International agreements, like the Outer Space Treaty, provide a framework for responsible exploration, but these frameworks will need to evolve as activity increases.

The search for life raises particularly profound ethical questions. If life is discovered on Mars, Europa, or another celestial body, humanity must decide how to interact with it—if at all. Protecting these ecosystems, even if microbial, is a priority for ensuring that exploration respects the intrinsic value of life.

Beyond its scientific and technological achievements, space exploration inspires a sense of wonder and possibility. It unites humanity in a shared pursuit of knowledge, transcending national and cultural boundaries. Each mission, whether robotic or crewed, represents a triumph of human ingenuity and determination.

The images of Earth taken from space—the "pale blue dot" and the "Earthrise"—remind us of our shared existence on a fragile planet. They underscore the importance of exploring space not as an escape from Earth's challenges but as a way to better understand and protect our home.

The Moon is the first step in a journey that will take humanity to the stars. Each milestone achieved on its surface prepares us for the challenges and opportunities of deeper exploration. As we look to Mars, asteroids, and beyond, the possibilities are as vast as the universe itself.

The future of space exploration is a testament to human creativity, resilience, and ambition. It is a story that is still being written, with each mission adding a new chapter. The Moon has shown us that we are capable of reaching beyond what was once thought possible. The next steps will take us further, into a future where the boundaries of exploration are limited only by our imagination.

As humanity embarks on a new era of space exploration, the Moon serves as a critical foundation for expanding our reach into the solar system. The next steps in space exploration will leverage the knowledge, technology, and infrastructure developed through lunar missions to advance toward Mars, the asteroid belt, and even interstellar travel. These endeavors will redefine what it means to explore, innovate, and collaborate on a global scale.

Mars stands as the next major milestone in humanity's journey beyond Earth. Unlike the Moon, Mars has an atmosphere—albeit thin—and signs of past liquid water, making it an intriguing candidate for studying planetary evolution and the potential for life. Robotic missions have already mapped much of its surface, revealing a landscape rich in geological diversity and scientific opportunities.

The lessons learned from lunar exploration are directly applicable to Mars. Technologies for extracting and using local resources, building habitats, and sustaining human life in harsh environments are all being refined on the Moon. These advancements will be critical for overcoming the challenges of Mars, which include greater distance from Earth, longer communication delays, and a more complex environment.

Mars missions will likely rely on the Moon as a staging ground. Spacecraft can be assembled and refueled in the Moon's low-gravity environment, reducing the energy required for interplanetary travel. This approach makes the Moon not just a stepping stone but a key logistical hub for future exploration.

Before humans set foot on new worlds, robotic missions will continue to lead the way. These missions are essential for scouting landing sites, mapping resources, and testing technologies. On Mars, for example, the Perseverance rover is already conducting experiments to produce oxygen from the planet's atmosphere, paving the way for future human missions.

Similar efforts are planned for asteroids and the icy moons of the outer planets. Robotic probes will drill into subsurface oceans, collect samples, and search for signs of life. These missions expand the scope of exploration, reaching places that are currently beyond human capability while informing the design of future crewed missions.

Asteroids represent a wealth of untapped resources, from water and metals to rare minerals used in advanced technologies. The Moon's proximity to the asteroid belt makes it an ideal base for launching missions to these small, resource-rich bodies. Extracting resources from asteroids could support both space exploration and Earth-based industries, reducing the environmental impact of terrestrial mining.

Mining operations on asteroids will rely on technologies developed for lunar mining, including robotic systems and advanced processing techniques. The ability to harness these resources not only supports exploration but also lays the foundation for a space-based economy that could drive innovation and growth for decades to come.

The outer planets and their moons are among the most intriguing targets in the search for extraterrestrial life. Moons like Europa, Enceladus, and Titan are believed to harbor subsurface oceans or other environments capable of supporting life. Robotic missions are already being planned to explore these worlds, with a focus on detecting biosignatures and understanding their unique ecosystems.

The exploration of the outer solar system will push the boundaries of current technology. Nuclear-powered spacecraft, long-duration life support systems, and advanced communication networks are all critical for reaching and studying these distant destinations. Each mission brings humanity closer to answering one of its most profound questions: Are we alone in the universe?

Advancements in propulsion technology will be key to the future of space exploration. Traditional chemical rockets, while effective for reaching the Moon and low Earth orbit, are less efficient for interplanetary missions. New propulsion systems, such as nuclear thermal propulsion, electric ion drives, and solar sails, promise faster and more efficient travel.

Nuclear propulsion, in particular, offers significant advantages for missions to Mars and beyond. By using nuclear reactors to heat propellant, these systems can achieve higher speeds and reduce travel times. Electric propulsion, which uses magnetic fields to accelerate charged particles, is another promising option for long-duration missions.

For interstellar travel, concepts like antimatter engines and warp drives remain speculative but are inspiring research into the physics of faster-than-light travel. While these technologies are far from realization, they represent humanity's long-term aspirations to explore the galaxy.

The next phase of space exploration is a collaborative effort that transcends national and cultural boundaries. International partnerships are becoming the norm, as countries pool resources and expertise to achieve shared goals. The Artemis program, which aims to return humans to the Moon and establish a sustainable presence, includes contributions from numerous nations and private companies.

Private industry plays an increasingly central role in exploration. Companies like SpaceX, Blue Origin, and Boeing are developing the infrastructure needed to support ambitious missions, from reusable rockets to orbital habitats. These partnerships between government and industry are accelerating progress, lowering costs, and opening space to new participants.

As humanity ventures further into the cosmos, ethical considerations become paramount. Protecting planetary environments, avoiding contamination, and ensuring equitable access to resources are all critical issues. International agreements, like the Outer Space Treaty, provide a foundation for responsible exploration, but they will need to evolve as activity increases.

The search for life raises profound ethical questions. If microbial life is discovered on Mars or an icy moon, humanity must decide how to interact with it—if at all. Protecting these ecosystems while advancing scientific knowledge is a delicate balance that requires careful thought and collaboration.

The future of space exploration is as much about inspiration as it is about discovery. Each mission, whether robotic or human, captures the imagination of millions, demonstrating what humanity can achieve through innovation and determination. Space exploration unites people across the globe, offering a shared sense of purpose and possibility.

The iconic images of Earth from space, the exploration of distant planets, and the pursuit of answers to fundamental questions about life and the universe remind us of our place in the cosmos. They challenge us to think beyond our immediate concerns and consider the broader trajectory of humanity.

The Moon has shown us what is possible, but it is only the beginning. The next steps—Mars, asteroids, the outer planets, and beyond—will push the boundaries of science and exploration. Each success builds on the achievements of the past, propelling humanity toward a future where the stars are within our reach.

The journey ahead is filled with challenges, but it is also rich with opportunity. By working together, embracing innovation, and respecting the cosmos, humanity can forge a future that honors its adventurous spirit and ensures its place among the stars.

Humanity's future in space is poised to enter a phase of unparalleled ambition and innovation. With the Moon serving as a foundation for exploration, the next decades will witness efforts to expand human presence across the solar system. From pioneering missions to Mars to exploring the icy worlds of the outer planets, these endeavors represent a blend of scientific inquiry, technological mastery, and the indomitable human spirit.

While initial missions to Mars will focus on exploration, the long-term vision involves the establishment of permanent colonies. Colonization requires solving challenges related to sustainability, such as producing food, water, and energy in an environment that lacks these resources. Technologies being tested on the Moon, such as in-situ resource utilization (ISRU), will play a critical role in enabling Mars colonization.

Mars habitats must be designed to withstand the planet's harsh conditions, including its thin atmosphere, cold temperatures, and high radiation levels. Dome-like structures, subterranean habitats, and inflatable modules covered in Martian regolith for radiation shielding are among the concepts being explored. These habitats will integrate life support systems capable of recycling air and water, ensuring long-term habitability.

Agriculture on Mars is a particularly challenging yet essential component of colonization. Researchers are experimenting with growing crops in simulated Martian soil, incorporating techniques like hydroponics and aeroponics to maximize efficiency. Advances in genetic engineering could produce crops specifically suited for Martian conditions, ensuring reliable food sources for settlers.

Asteroids are not only rich in scientific knowledge but also hold immense economic potential. These small bodies contain valuable metals, water, and organic compounds that can support both exploration and industrial activities. Mining asteroids for resources such as platinum-group metals and water could fuel a new era of space-based industry, reducing reliance on Earth's finite resources.

Robotic missions will pioneer asteroid mining, using automated systems to identify, extract, and process materials. The technologies developed for lunar and Martian mining will be adapted for asteroid operations, ensuring efficiency and safety. Some concepts even envision relocating smaller asteroids into orbits around the Moon or Earth, making them more accessible for mining operations.

Asteroids also serve as natural laboratories for studying the origins of the solar system. By analyzing their compositions, scientists can uncover clues about the processes that led to the formation of planets and the delivery of water and organic molecules to Earth.

The outer solar system is home to some of the most intriguing worlds, particularly the icy moons of Jupiter and Saturn. Europa, Ganymede, Enceladus, and Titan are believed to harbor environments that could potentially support life. Subsurface oceans beneath their icy crusts are kept warm by tidal heating, creating conditions similar to those that may have existed on early Earth.

Robotic missions will be the first to explore these moons in detail. Europa Clipper, for example, is set to conduct multiple flybys of Europa, studying its surface and assessing the habitability of its ocean. Future missions may include landers equipped with drills to penetrate the ice and submersibles to explore the oceans below.

The exploration of these moons represents one of humanity's most ambitious scientific goals. The discovery of life—or even the conditions for life—on another world would profoundly change our understanding of biology, the origins of life, and humanity's place in the universe.

While the solar system remains the primary focus of exploration, efforts are underway to lay the groundwork for interstellar missions. Concepts such as Breakthrough Starshot, which aims to send small, light-powered probes to the Alpha Centauri system, demonstrate the feasibility of exploring beyond our solar neighborhood.

Advances in propulsion technology are critical for interstellar travel. Nuclear propulsion, solar sails, and even speculative ideas like warp drives are being studied for their potential to enable missions to other star systems. While interstellar travel is still decades or centuries away, these efforts represent humanity's long-term vision for exploring the cosmos.

The ultimate goal of space exploration is to establish a multiplanetary civilization. This vision extends beyond survival, encompassing the creation of thriving societies on other worlds. Such a civilization would leverage the unique characteristics of each celestial body—Mars for its resources and habitability potential, asteroids for their metals and volatiles, and the Moon for its proximity and infrastructure.

A multiplanetary civilization also diversifies humanity's presence, reducing the risks associated with being confined to a single planet. This diversification is not only practical but also symbolic, representing humanity's resilience and adaptability in the face of challenges.

As space exploration ventures into increasingly complex and distant environments, artificial intelligence (AI) and robotics will play a central role. Autonomous systems are already being used to explore the Moon, Mars, and asteroids, and their capabilities are rapidly advancing. AI enables robots to make decisions, adapt to changing conditions, and perform tasks with minimal human intervention.

Future missions will rely on AI to manage spacecraft, analyze data, and support human crews. On Mars, for instance, AI could oversee agricultural systems, monitor habitat conditions, and assist with scientific research. These technologies are essential for extending humanity's reach into the cosmos while ensuring efficiency and safety.

The challenges of space exploration drive technological advancements that benefit life on Earth. From energy systems and medical devices to materials science and environmental monitoring, the innovations developed for space missions often find applications in everyday life. These spin-offs demonstrate the value of investing in exploration, not just for its direct outcomes but for its broader impact on society.

Space exploration also fosters international collaboration and cultural exchange. The shared pursuit of ambitious goals unites nations, encourages cooperation, and inspires generations to think beyond borders. This spirit of exploration transcends politics and competition, highlighting humanity's collective potential.

The future of space exploration is a story of possibility and progress. It is a journey that began with the first human gazes toward the stars and continues with every mission to the Moon, Mars, and beyond. Each step forward represents a leap toward understanding our universe and our place within it.

As humanity prepares for the challenges ahead, the possibilities are as vast as the cosmos itself. By embracing innovation, collaboration, and sustainability, we can ensure that space exploration remains a source of inspiration and discovery for generations to come.

Chapter 10: Ethical Frontiers in Exploration

As humanity ventures deeper into space, the challenges of exploration are not only technical and logistical but also ethical. The vastness of space does not exempt us from the moral questions that arise when engaging with new frontiers. From the treatment of extraterrestrial environments to the equitable distribution of space resources, the future of space exploration demands careful thought and responsible action. Ethical considerations will shape how we approach these endeavors, ensuring that exploration benefits humanity while respecting the integrity of the cosmos.

One of the foremost ethical challenges is the preservation of extraterrestrial environments. Celestial bodies like the Moon, Mars, and Europa are pristine scientific records of the solar system's history. Their surfaces and subsurfaces contain invaluable clues about planetary formation, the potential for life, and cosmic evolution. Altering these environments through human activity risks losing critical data and disrupting natural processes.

Planetary protection policies aim to prevent contamination of these worlds. These guidelines, established by organizations like NASA and the European Space Agency (ESA), require strict sterilization of spacecraft to ensure that Earth microbes do not inadvertently colonize other planets. Similarly, if life is discovered elsewhere, it becomes essential to prevent backward contamination, where extraterrestrial organisms might pose risks to Earth's ecosystems.

Beyond microbial contamination, large-scale human activities, such as mining or habitat construction, must be carefully managed to minimize environmental impact. Questions about how much alteration is acceptable and who decides these limits are central to ethical space exploration.

The Moon and asteroids are rich in resources, including water ice, metals, and rare elements. As nations and private companies race to extract these materials, ethical questions about ownership and access become increasingly urgent. Who owns these resources, and how should they be distributed? What frameworks are needed to ensure fair access for all nations, especially those without robust space programs?

Current international agreements, such as the Outer Space Treaty of 1967, establish space as a global commons, stating that no nation can claim sovereignty over celestial bodies. However, as commercial activities increase, the need for more specific regulations is evident. Initiatives like the Artemis Accords attempt to address these gaps by promoting peaceful exploration and resource sharing, but global consensus remains elusive.

Equitable access also extends to the benefits of space exploration. How can advancements made through exploration improve life on Earth, particularly for marginalized communities? Ethical exploration requires ensuring that the gains from space—technological, economic, and scientific—are distributed broadly and inclusively.

The search for life beyond Earth raises profound ethical questions. If we discover microbial life on Mars, Europa, or another world, what is our responsibility toward it? Should extraterrestrial life be treated as a resource to study and utilize, or does it have intrinsic value deserving of protection? These questions echo debates about conservation and animal rights on Earth, but in a context that is entirely unprecedented.

The principle of non-interference is often proposed as a guideline, suggesting that humanity should avoid disrupting extraterrestrial ecosystems. However, practical challenges arise when balancing scientific investigation with preservation. For instance, drilling into Europa's icy crust to access its subsurface ocean might irreversibly alter its environment, even if done with the utmost care.

Establishing ethical guidelines for interacting with extraterrestrial life will require input from scientists, ethicists, policymakers, and the global community. These guidelines must balance the desire for knowledge with the responsibility to protect and respect alien ecosystems.

The prospect of colonizing the Moon, Mars, and other celestial bodies introduces a host of ethical dilemmas. Colonization, historically associated with exploitation and inequality on Earth, must not repeat these patterns in space. Ensuring that space settlements are inclusive, equitable, and sustainable is critical to avoiding the mistakes of the past.

The allocation of land and resources on other worlds must be governed by transparent and fair systems. Additionally, the rights and well-being of settlers must be protected. What legal frameworks will govern space colonies? How will disputes be resolved, and who will hold authority over these new communities? These questions highlight the need for robust policies that prioritize human rights and shared values.

The growing involvement of private companies in space exploration has accelerated progress but also raises ethical concerns about commercialization. As companies invest in mining, tourism, and other ventures, the balance between profit and public good becomes a key issue. How can the benefits of private innovation be harnessed while ensuring that exploration serves humanity as a whole?

Transparency and accountability are essential in addressing these concerns. Private ventures must operate within ethical frameworks that prioritize sustainability, equitable access, and respect for celestial environments. Partnerships between governments and private companies can help align commercial activities with broader societal goals.

Space exploration is inherently forward-looking, but it must also consider the long-term consequences of human activity in space. Decisions made today will shape the opportunities available to future generations. Overextraction of resources, environmental degradation, and geopolitical conflicts could limit humanity's ability to explore and thrive in space.

Adopting a stewardship mindset is crucial for ensuring that space exploration remains a sustainable and inclusive endeavor. This approach emphasizes the careful management of resources, the preservation of environments, and the avoidance of actions that could jeopardize future exploration.

Exploration is not solely a scientific and technological pursuit—it is also deeply cultural and philosophical. Space inspires profound questions about humanity's place in the universe, our relationship with the cosmos, and the meaning of life itself. Ethical exploration requires respecting the cultural and spiritual significance that space holds for different communities and perspectives.

For instance, the Moon has symbolic importance in many cultures, from ancient myths to modern traditions. Recognizing and honoring these connections enriches the global narrative of space exploration, fostering a sense of shared heritage and purpose.

Ethical space exploration requires collaboration on a global scale. No single nation or entity should dominate the narrative or direction of exploration. By working together, humanity can address the complex ethical challenges of space and ensure that its benefits are shared equitably.

Organizations like the United Nations, space agencies, and international consortia play a critical role in shaping policies and fostering dialogue. Public engagement is equally important, as exploration affects not only scientists and policymakers but all of humanity.

As humanity takes its first steps into the cosmos, the ethical dimensions of exploration must guide every decision. Protecting celestial environments, respecting the rights of all stakeholders, and prioritizing sustainability are essential for building a legacy that future generations can be proud of. Space exploration is not just about reaching new worlds—it is about how we choose to explore them and the values we bring with us.

The ethical dimensions of space exploration extend beyond immediate concerns of planetary protection and resource management. As humanity prepares for deeper ventures into the cosmos, the challenges of governance, cultural inclusion, and intergenerational equity demand careful consideration. How we navigate these issues will shape the legacy of our exploration, ensuring that it aligns with humanity's highest aspirations.

The governance of space activities remains one of the most complex ethical challenges. Existing treaties, such as the Outer Space Treaty of 1967, provide a foundational framework, declaring space a global commons to be used for the benefit of all humanity. However, these agreements were created in an era when space exploration was limited to a few nations and primarily governmental in nature. The emergence of private companies and the increasing number of spacefaring nations have introduced new dynamics that these treaties do not fully address.

Establishing governance mechanisms that ensure accountability, transparency, and fairness is essential. This includes defining rules for resource extraction, regulating the environmental impact of space activities, and ensuring that commercial interests do not overshadow broader human goals. New agreements must also address the potential for militarization and conflict in space, promoting peaceful collaboration over competition.

The Moon, Mars, and other celestial bodies hold not only scientific but also cultural and historical significance. The Moon, for instance, has been a source of inspiration for cultures worldwide, representing themes of mystery, renewal, and connection. Protecting the cultural legacy of these celestial bodies is as important as preserving their physical environments.

Sites of historical importance, such as the Apollo landing sites, are particularly significant. These locations represent milestones in human achievement and should be treated with the same care as archaeological sites on Earth. International agreements are needed to ensure that these sites are preserved for future generations, avoiding disruption or commercialization.

Space exploration has the potential to unite humanity, but it also risks reinforcing existing inequalities if not approached with care. Ensuring that all nations and communities have a voice in shaping the future of space is critical for fostering inclusion and equity. This includes providing access to the benefits of space exploration, such as technological advancements and economic opportunities, regardless of a nation's wealth or technological capabilities.

Diversity in the space workforce is equally important. Expanding opportunities for underrepresented groups, including women, minorities, and individuals from developing countries, enriches the perspectives and innovations that drive exploration. International programs like the United Nations' Office for Outer Space Affairs (UNOOSA) play a vital role in promoting inclusivity and capacity-building, ensuring that space exploration reflects the diversity of humanity.

Artificial intelligence (AI) and autonomous systems are integral to modern space exploration, performing tasks that are too dangerous or complex for humans. However, the increasing reliance on AI raises ethical questions about decision-making and accountability. For example, autonomous systems may be tasked with exploring fragile environments, and their actions could inadvertently cause harm.

Ensuring that AI systems are aligned with human values and guided by clear ethical principles is essential. This includes establishing frameworks for accountability when autonomous systems operate independently or make decisions in real-time. These considerations extend beyond space exploration, offering insights into the broader integration of AI in society.

As humanity taps into the resources of the Moon, asteroids, and other celestial bodies, the question of resource scarcity becomes increasingly relevant. While the resources of space appear vast, they are not infinite. Responsible utilization requires balancing immediate needs with long-term sustainability, ensuring that future generations have access to the same opportunities.

Establishing quotas, conservation strategies, and recycling systems for extracted materials can help mitigate overexploitation. Additionally, mechanisms for redistributing the benefits of space resources to underserved communities on Earth can promote ethical utilization. For example, revenues from space mining could be reinvested in global development projects, advancing education, healthcare, and sustainability.

The precautionary principle—avoiding actions with unknown or potentially harmful consequences—is a cornerstone of ethical space exploration. This principle is particularly relevant in the context of planetary preservation. As humanity considers large-scale activities on the Moon and Mars, such as mining, terraforming, or colonization, it is crucial to understand the potential impacts before proceeding.

For Mars, the possibility of existing microbial life adds an additional layer of complexity. Contamination from Earth-based organisms could irreversibly alter the planet's environment and obscure scientific findings. Adopting stringent sterilization protocols and limiting human activity in high-priority research areas are essential steps in protecting Mars' potential ecosystems.

Space exploration is inherently a forward-looking endeavor, but its benefits and consequences will unfold over decades and centuries. Decisions made today must account for their impact on future generations, ensuring that exploration remains sustainable and inclusive. This principle of intergenerational equity emphasizes the responsibility to manage space resources, environments, and opportunities with long-term stewardship in mind.

Education plays a vital role in fostering this vision. By inspiring young people to pursue careers in science, technology, and exploration, humanity can ensure that future generations are equipped to continue the journey. Programs that engage students from diverse backgrounds in space-related disciplines contribute to a legacy of innovation and collaboration.

Ethical space exploration requires a unified vision that integrates science, technology, culture, and governance. It calls for a commitment to shared values, such as sustainability, inclusivity, and respect for the cosmos. This vision transcends national and commercial interests, focusing instead on the collective good of humanity and the preservation of space as a frontier for discovery.

Public engagement is a critical component of this vision. Transparent communication about the goals, risks, and benefits of exploration fosters trust and ensures that decisions reflect the will and interests of global citizens. Artistic and cultural expressions inspired by space exploration further enrich this dialogue, creating a shared narrative that resonates across generations.

As humanity steps into the next phase of exploration, the ethical challenges ahead are as vast as the cosmos itself. From preserving the integrity of celestial bodies to ensuring equitable access to resources, these challenges require innovative solutions and unwavering commitment. The choices we make today will define the character of our exploration and the legacy we leave for future generations.

By prioritizing ethics in every aspect of space exploration, humanity can chart a path that honors the spirit of discovery while respecting the values that unite us. In doing so, we ensure that the pursuit of the stars remains not only a testament to human ingenuity but also a reflection of our shared humanity.

Epilogue: A Vision Beyond the Horizon

As humanity gazes upward, the Moon remains a symbol of what we have achieved and a beacon for what lies ahead. It has been a silent witness to our first tentative steps into the cosmos, a testing ground for our ingenuity, and a gateway to infinite possibilities. Each discovery made on its surface and each innovation born from its exploration brings us closer to realizing our collective dream of reaching beyond the confines of Earth.

The journey to the Moon, and beyond, is not just a story of science and technology—it is a story of resilience, imagination, and the enduring human spirit. It reflects the curiosity that has driven humanity to explore, the creativity that has solved insurmountable challenges, and the unity that comes from pursuing goals larger than any individual or nation.

The Moon has shown us that we are capable of extraordinary achievements when we work together. It has taught us to respect the environments we explore, to value the shared heritage of our world, and to plan with foresight and responsibility for the generations to come. These lessons are as vital as the technologies and scientific knowledge we gain, for they remind us that exploration is not just about where we go, but how and why we go.

Looking forward, the Moon is not an endpoint but a beginning. It is a stepping stone to Mars, the asteroids, the icy moons of the outer planets, and perhaps even the stars themselves. It challenges us to think beyond immediate horizons and to envision a future where humanity is a multiplanetary species, thriving in the vastness of space.

This future is not without challenges. Ethical dilemmas, environmental responsibilities, and questions of equity and inclusion demand our attention and care. Yet, these challenges also inspire us to grow, to innovate, and to ensure that our exploration reflects the best of who we are.

As we continue to write the next chapters of this story, the Moon will remain a constant companion and a silent partner in our journey. It will inspire poets, guide scientists, and challenge engineers, serving as a reminder of both our place in the universe and our potential to transcend it.

The Moon calls to us not just as a destination but as an invitation—to learn, to grow, to explore. By answering this call, we affirm that the human spirit is limitless, capable of reaching for the stars while remaining deeply connected to the world that gave us life.

The horizon beckons. The journey continues.

www.ingramcontent.com/pod-product-compliance
Lightning Source LLC
Chambersburg PA
CBHW071127240526
45465CB00024B/1436